SpringerBriefs in Computer Science

More information about this series at http://www.springer.com/series/10028

Anastasios Bezerianos · Andrei Dragomir
Panos Balomenos

Computational Methods for Processing and Analysis of Biological Pathways

Springer

Anastasios Bezerianos
University of Patras
Patras
Greece

and

Centre of Life Sciences
National University of Singapore
Singapore
Singapore

Andrei Dragomir
University of Houston
Houston, TX
USA

Panos Balomenos ⓘ
University of Patras
Patras
Greece

ISSN 2191-5768   ISSN 2191-5776   (electronic)
SpringerBriefs in Computer Science
ISBN 978-3-319-53867-9   ISBN 978-3-319-53868-6   (eBook)
DOI 10.1007/978-3-319-53868-6

Library of Congress Control Number: 2017932623

Printed on acid-free paper

This Springer imprint is published by Springer Nature
The registered company is Springer International Publishing AG
The registered company address is: Gewerbestrasse 11, 6330 Cham, Switzerland

Preface

Drug design is a complex and resource-demanding process, hinged on the inherent complexity of diseases, which arise from deregulated interactions between multiple genes. The reductionist approach, which has served the pharma industry for decades, has been producing declining R&D returns in recent years, leading not only to ever-increasing costs, but also to reduced efficacy, thus affecting the quality of healthcare. More than a decade ago, developing a new drug was associated with a cost of $890 million. In 2016, the estimate is a staggering $2870 million. These costs are driving the emergence of new technologies and methodologies which will eventually shape a phenomenal paradigm shift in the research and development of drugs.

Most disease-associated genes have a small influence by themselves. However, in the context of a molecular interaction network of cellular processes, these individual influences are combined in such a way that the resulting influence severely deregulates the network. In that regard, both the disease and the corresponding treatment introduce perturbations within the biological network, which should be assessed by means of a systems-level approach.

Pathway analysis is a thriving research area in systems biology, attempting to unravel the systemic effects of disease and drug-induced perturbations. Pathway analysis relies on the wealth of complex biological data produced by omics technologies. These technologies typically produce a list of differentially expressed genes between a control and a disease state. The data, however, have been removed from the biological context from which they were extracted. Pathway analysis methods attempt to rectify this by using prior biological knowledge pertaining to the structure and operation of biological pathways along with statistical, mathematical, and computational methods.

Towards the direction of understanding what pathway analysis can offer to both the experimentalist and the modeler, the reader is introduced in the first chapter to a general methodology which outlines common workflows shared by several methods, such as preprocessing of the omics data, choosing a null hypothesis, as well as gene- and pathway-level statistics. Furthermore, the evolution of pathway analysis methods is documented, beginning from simple overrepresentation analysis and leading to complex pathway-level, or even sub-pathway-level approaches. We continue in the second chapter with a comprehensive review covering pathway and sub-pathway based approaches involved in various aspects of drug design and discovery, an emerging area known as systems pharmacology. This chapter provides insights into how pathway analysis methods can be employed to elucidate drug mechanism of action, identify novel drug targets, increase treatment efficacy by identifying drugs or drug combinations, which modulate multiple targets, infer novel therapeutic indications for existing drugs, and predict drug side effects.

Subsequently, in the third chapter, we present an overview of pathway analysis methods developed to model the temporal aspects of drug- or disease-induced perturbations and extract relevant dynamic themes. In the fourth and final chapter, several state-of-the-art methods in pathway analysis are outlined, which address the important problem of identifying differentially expressed pathways and sub-pathways. We cover various aspects of the methodological arsenal of this area, as well as the evolution of tools developed for differential expression analysis.

In conclusion, the present work offers the reader a guided walkthrough to one of the most promising research areas in modern life sciences, enabling a deeper understanding of involved concepts and methodologies via an interdisciplinary view, focusing from well-established approaches to cutting-edge research.

Patras, Greece   Anastasios Bezerianos
Houston, USA   Andrei Dragomir
Patras, Greece   Panos Balomenos

Contents

1 **Introduction** 1
  1.1 Biological Networks 2
    1.1.1 Properties 2
    1.1.2 Categories of Biological Networks 3
  1.2 Pathway Analysis 4
    1.2.1 Methodology 5
    1.2.2 Evolution of Pathway Analysis Methods 6
  1.3 Systems Pharmacology 8
  References 9

2 **Networks and Pathways in Systems Pharmacology** 11
  2.1 Introduction 11
  2.2 Network- and Pathway/Sub-pathway-Based Characterization
    of Drugs Mechanism of Action 15
  2.3 Identification of New Drug Targets and Polypharmacology
    Applications 24
    2.3.1 Target Characterization and Identification Using Network
      Properties of Drug Targets 24
    2.3.2 Identification of Drug Targets Based on Integrative
      Network Approaches 26
    2.3.3 Network-Based Polypharmacology 28
  2.4 Network-Based Drug Repositioning 29
    2.4.1 Drug Repositioning Based on Molecular Profiles 30
    2.4.2 Drug Repositioning Based on Phenotypic Profiles 33
  2.5 Network-Based Side Effect Modeling and Prediction 35
    2.5.1 Approaches Based on Chemical Structure 36
    2.5.2 Approaches Based on Pathways/Sub-pathways 38
  2.6 Current Challenges and Future Considerations 39
  References 41

3   Time-Varying Methods for Pathway and Sub-pathway Analysis....   47
   3.1   Introduction   47
   3.2   Conversion of Pathway Databases to Graphs   50
   3.3   Time-Varying Pathway and Sub-pathway Extraction   54
      3.3.1   Linear Sub-pathway Extraction   54
      3.3.2   Nonlinear Sub-pathway Extraction   56
   3.4   Temporal Dynamics Scoring Schemes   57
      3.4.1   Approach Followed in CHRONOS...................   57
      3.4.2   Other Approaches................................   61
   3.5   Synthetic and Biological Data Analysis Results   65
   References...   67

**4   Identification of Differentially Expressed Pathways
and Sub-pathways**......................................   69
   4.1   Introduction   70
   4.2   Approaches Based on Gene Sets   70
   4.3   Approaches Based on Pathway Topology...................   72
   4.4   Conversion of Pathway Databases Information to Graphs   79
   4.5   Gene and Pathway-Level Statistics of Differential Expression   80
      4.5.1   Methods Using Experimental Data for Sub-pathway
Identification....................................   80
      4.5.2   Methods Using Experimental Data for Evaluating
the Statistical Significance of Sub-pathways   81
   4.6   DEsubs: A Flexible Tool for Identification of Differentially
Expressed Sub-pathways.................................   82
   References...   85

Chapter 1
Introduction

Abstract This chapter presents an overview of the topics discussed throughout the book. The use of network representations for describing biological processes from a systems biology perspective is first highlighted. Subsequently, various types of biological networks are scrutinized and a synopsis of pathway analysis methods is provided. The evolution of pathway analysis approaches is then presented, and various methodological aspects involved are introduced. The chapter concludes by presenting several aspects related to one of the main application areas of pathway analysis: systems pharmacology, an emerging interdisciplinary and translational field at the confluence of systems biology and pharmacology. The relevance of a pathway-level analysis framework in the drug design and discovery pipeline is also discussed, especially for the identification of novel drug targets, drug repositioning, as well as drug safety and side effect prediction.

Keywords Biological networks · Pathway analysis · Sub-pathways · Systems biology · Systems pharmacology

Drug design is a complex, costly, and time-consuming process suffering from costs incurred by the intrinsic complexity of diseases, including myriads of multilevel interactions between a large number of gene-products. The traditional reductionist's approach, which entails the elucidation of disease mechanism and drug mode of action through the study of few genes reached its zenith in the 1990s and has been producing diminishing returns since the beginning of the 2000s. The most staggering realization of that fact is that the cost of developing a new drug has skyrocketed from $1 billion to $2.8 billion in the past fifteen years (DiMasi et al. 2003, 2016).

Most disease-related genes have small effects with a combined significant impact (Petretto et al. 2007). Since these genes operate from a complex multilevel network, the relevance of any biomarker is entwined with the very network it is a part of (Gustafsson et al. 2014). Moreover, the effect of a treatment is rarely localized since it may be propagated throughout the biological network, thus changing the picture of the organism in unforeseen ways (Hidalgo et al. 2009).

© The Author(s) 2017
A. Bezerianos et al., *Computational Methods for Processing and Analysis of Biological Pathways*, SpringerBriefs in Computer Science, DOI 10.1007/978-3-319-53868-6_1

Systems Biology is an emerging research field studying biological systems by introducing systematically systemic perturbations, monitoring the system's response in multiple levels and consolidating the responses by means of mathematical and statistical models. It transcends the limitations of the reductionist's approach and incorporates information ranging from the DNA up to populations. A more detailed framework, according to (Ideker et al. 2001) is as follows:

(i) Creation of a model describing the system in sufficient detail by means of available genetic and biochemical data. Such a model can explain the system-wide interactions under observation, and at the same time anticipate its responses, given specific new perturbations introduced to the system.

(ii) Introduction of systematic genetic, biochemical, or environmental perturbations in the system, and monitoring of changes, such as differential expression of genes.

(iii) Update of the model to account for novel experimental results, in case it fails to explain them and so on. Thus, the model drives the design of the experiments, where new discoveries in turn drive the update of the model itself, driving the next series of experiments.

Over the course of the past two decades, the cost of DNA-sequencing has dropped significantly. This cost, at the time of the publication of the first draft of the human genome in 2001 was in the vicinity of $95 million and is nowadays in the vicinity of $1000 (Wetterstrand 2016). This evolution may eventually enable the design of personalized drugs, targeted toward stratified groups of patients, according to both genetic and environmental criteria, therefore addressing the low productivity of the pharma industry in the most meaningful of ways.

1.1 Biological Networks

1.1.1 Properties

Biological networks are comprised of the genes encoding proteins that are responsible for the structure and function of an organism, as well as a level of interactions that regulate gene expression. Other micro-molecules and macro-molecules are also part of those networks. Biological networks have a series of interesting properties, whose elucidation promoted Systems Biology (Barabási and Réka 1999).

Biological networks are scale-free networks. In a scale-free network, the probability of a node's degree follows the power-law distribution. It exhibits preferential connectivity, since the probability that a new edge is added to the network is proportional to the node's degree. Thus, few nodes tend to have high degrees (hubs), with the majority of the nodes exhibiting limited connectivity. The hubs are rarely interconnected but are connected with nodes belonging in few interactions. The mechanism that seems to drive this property is gene duplication (Barabási and

Réka 1999). When a cell is divided, some gene may appear in two copies in the descendant's genome. In this way, a new node is added within the network of gene/protein interactions. The new protein has the same structure as the original one and therefore interacts with the same proteins, and all of its interaction partners gain a new interaction from the duplicated gene. Thus, proteins sharing a large number of interactions with other proteins tend to gain new connections more frequently (Barabási and Réka 1999).

Biological networks exhibit modules and motifs. A module is characterized by the existence of several interactions between its members, compared to the ones with members outside the module. Within those modules there exist subgraphs which are specific to the identity of the biological network. Those subgraphs which appear more frequently in the biological network than in a randomized version with the same number of nodes and degree distribution are called motifs (Milo et al. 2002).

Biological networks also exhibit the ultra small-world effect in addition to a complex network's small-world effect, where any node can be reached from another node within few steps (Milgram 1967). In this case, scale-free networks have a much smaller average path length (Cohen and Havlin 2003). This property depicts the fact that small local perturbations have the potential of being transmitted throughout the network very quickly.

1.1.2 Categories of Biological Networks

Protein Interaction Networks (PIN)

The nodes of a PIN represent proteins and the edges the interaction between two proteins. The edges are undirected, since if protein A interacts with protein B, then the opposite is also true. The edge can also represent a brief interaction which modifies a protein. Such interactions represent the dynamic part of the network. Example of such interactions include protein kinases, which add a phosphate group on a target-protein, and karyopherins which transport molecules between the cytoplasm and the nucleus (Meyers 2009).

Metabolic Networks (MN)

Metabolic networks are comprised of metabolic pathways, each being a sequence of biochemical reactions within a cell, catalyzed by enzymes, with the product of one reaction either being used in the cell or acting as a substrate to a successive reaction. The nodes of the network represent the products and the edges the reactions between them.

Gene Regulatory Networks (GRN)

The main mechanism of gene expression regulation entails the regulation of transcription through proteins called transcription factors. A transcription factor is bound on specific DNA sequences, changing the concentration of RNA-polymerase which is responsible for transcribing the DNA sequence to an RNA sequence. A single transcription factor may regulate multiple genes and a single gene may be regulated by several transcription factors. The nodes represent genes and transcription factors, and the edges directed interactions from a transcription factor to a gene whose expression is regulated.

Signaling Networks (SN)

A signaling network controls basic cellular processes and is comprised of signaling pathways, each representing the flow of biological signal when a receptor of the cellular membrane is activated. The nodes of the network represent genes and the proteins they encode, which are part of the biological signal transduction through chemical transformation they are involved in. One example of such transformations is posttranslational modifications of proteins, where a functional group, such as phosphate or methyl group, is added to the protein, resulting in the change of its signaling function (phosphorylation, methylation). Another example is the formation of stable protein complexes (association) or the removal of a protein from such a complex (dissociation). Each gene may be a part of several pathways, therefore has the capacity to propagate the signal in multiple pathways. Each pathway usually begins with the gene encoding a cellular membrane ligand and ends with the change in the intracellular concentrations of signaling micro-molecules. Signaling networks are the most complex of networks, since they represent the biological signal transduction within the cell, taking into consideration both the type of interaction, as well as their time frame (Meyers 2009).

1.2 Pathway Analysis

Pathway analysis is a family of Systems Biology methods used to extract knowledge from the data produced by high-throughput sequencing technologies, by creating a model which summarizes and describes the underlying biological processes. These technologies typically produce a list of differentially expressed genes, between a case of interest, for example disease or drug perturbations and a control. Since this output is stripped from its biological context, pathway analysis techniques utilize prior biological knowledge with statistical, mathematical, and computational methods in order to derive a suitable model linking the data with the complex biological processes which they describe. The input of a pathway analysis method is the omics data produced by high-throughput technologies in omics studies, such as: (i) genomics, the study of the structure and function of an organism's genome, (ii) proteomics, the study of the structure and function of the proteins of an organism, (iii) transcriptomics, the study of RNA molecules transcribed in a cell, and (iv) metabolomics, the study of the products of enzyme biochemical reactions. Next, an indicative methodology in pathway analysis will be outlined (Goeman and Bühlmann 2007).

1.2.1 Methodology

Preprocessing

The first stage of the methodology is the preprocessing of the biological data which have been produced by an omics study, as well as the data available from a source of prior knowledge, such as a biological database. For the omics data, a usual preprocessing includes: (i) normalization for all samples, so that expression values from different experiments are comparable, (ii) removal or extrapolation of missing values, (iii) conversion from manufacturer annotation to a more neutral annotation, such as Entrez Id, (iv) removal of samples with no annotation matches (Hung et al. 2011). For the pathway data, the selection of one or more biological databases, such as protein–protein interactions or signaling pathways databases, is made not only by considering the relevance of each database with the aim of the method, but also the degree in which the prior knowledge covers the genes/transcripts of the experiment. No database fully covers the whole of an organism genome/transcriptome, and also, different databases may offer conflicting information regarding interactions or pathway structure.

Null Hypothesis Selection

The analysis phase of the methodology entails the identification of pathways which are perturbed in response to a biological stimulus, such as a disease or a drug perturbation. The perturbed pathways as a rule have to be statistically significant, in the sense that the responses are not be found by chance. The statistical significance is usually assessed using a P-value of less than 0.05, which is additionally corrected for multiple hypothesis testing in order to obtain a Q-value. The usual choice for a null hypothesis H_0 is that the genes belonging in a pathway have no relation to the observed phenotype (Ackermann and Korbinian 2009; Goeman and Bühlmann 2007). The correlation of the genes within the pathway (or any gene-set) with the phenotype is compared with the correlation of those genes with random phenotypes. The null hypothesis is that the genes within the pathway are not differentially expressed (association null hypothesis). This hypothesis has a clear biological meaning, with the P-value corresponding to the theoretical replication of the same experimental procedure giving new phenotypes. A significant P-value shows that the results are not random, while a significant Q-value shows that no or at least few false positives have been introduced as a result of multiple hypothesis testing (Ackermann and Korbinian 2009).

Gene-Level Statistic Selection

The next step is the identification of the differentially expressed genes between treatment and control samples. Some simple and yet powerful statistics are t-statistics, fold-change, Pearson correlation coefficient, Kolmogorov–Smirnov test. These statistics may also be transformed by considering absolute values, squared values, binary transformation, P-values, Q-values, or posterior Bayesian probabilities (Ackermann and Korbinian 2009; Khatri et al. 2012).

Gene-Set Level Statistic Selection

After having selected a gene-level statistic, it is transformed to a gene-set level statistic by considering the sum or the average of the transformed gene-level statistics, Kolmogorov–Smirnov statistics, Wilcoxon rank sum, or false discovery rate (FDR). The results produced by the gene-set level statistic will be evaluated as to their statistical significance according to the selected null hypothesis. In the case of the association null hypothesis, the phenotypes under observation are repeatedly perturbed and a new statistic is calculated (Ackermann and Korbinian 2009; Khatri et al. 2012).

1.2.2   Evolution of Pathway Analysis Methods

Pathway analysis methods can be categorized in four generations according to the hypotheses employed and the direction of the analysis (García-Campos et al. 2015; Khatri et al. 2012).

Overrepresentation Approach

The basic hypothesis in an overrepresentation analysis (ORA) is that a statistically significant pathway contains more differentially expressed genes than the ones that would appear by chance. Such methods tend to follow the general methodology outlined earlier very closely. As the first evolution of pathway analysis approaches, they had significant advantages compared to methods that do not utilize prior biological knowledge. The omics data are placed in a biologically relevant context which promotes the development of a model describing the changes in the complex biological processes behind the differential expression and the perturbed pheno-types. These methods are also simple and computationally inexpensive. However, they do have several restrictions: (i) Due to the fact that the selection of differen-tially expressed genes is performed based on a threshold, several potentially rele-vant genes close to the threshold may not be included in the list. (ii) While the pathways themselves have specific connections and direction, these methods view them as gene sets. Thus, if the edges of the network were to be randomly shuffled, the results of the method would be identical, which may compromise the signifi-cance of the results. (iii) Gene-level statistics tend to treat all genes equally ignoring how they interact with their neighbors. These restrictions paved the way for the next generations of approaches.

Functional Class Scoring

The basic hypothesis in a functional class scoring analysis (FCS) is that the pathway status is not only affected by significant changes in gene expression, but also by smaller changes with a combined significant contribution. Such methods also follow the general methodology outlined earlier, with the difference that the gene-set statistic is actually a pathway-level statistic, thus it should take into consideration not only information concerning the genes themselves, such as fold-change, but also the types of interactions they are parts of. One main benefit of such methods is that by taking into consideration the interactions between genes, they can identify combined changes in gene expression affecting the status of the pathway. This generation of approaches, however, does not fully address the previous generation's restrictions. (i) While they do take into account interactions between genes, all genes have the same contribution in the gene-level statistic, disregarding available prior knowledge. (ii) They do not take the pathway topology itself into consideration, but merely individual interactions. The third evolution, however, did address these issues.

Pathway Topology Analysis

The basic hypothesis in a pathway topology analysis is that the interactions contained in the pathway topology are important in studying the correlation of changes that occur between the different parts of a pathway. Such methods also follow the general methodology outlined earlier, however when calculating the gene-level statistic, the whole of the pathway topology is taken into consideration. The main benefit is that pathway topology provides information about the type of interaction between its members, which allows the assignment of different weights on each gene according to its changes and contribution to the pathway itself. The main restriction is that connections between the pathways themselves are not taken into account. They analyze each pathway independently ignoring interactions and overlaps between distinct pathways. If for example some of the differentially expressed genes in a pathway belong to another pathway as well, that pathway may be falsely identified to be correlated with the observed phenotype (Barabási and Réka 1999).

Sub-pathway Analysis

The basic hypothesis in a sub-pathway analysis is that specific biological processes can be better explained not by the perturbations in pathways, but rather specific sub-pathways within each pathway, which may be shared between them. These sub-pathways may exhibit the same role in different pathways (Chen et al. 2011). These approaches tend not to accurately be described by the general methodology. A defining part of each method is how it defines a sub-pathway, based on the kind of biological process it focuses on.

   Sub-pathways identify regions of interest within pathways which are perturbed as a result of a disease or a drug treatment. These approaches represent the pinnacle of pathway analysis methods: earlier generation approaches zoomed out of the gene level to encompass as much a systemic information as possible. Sub-pathway-level approaches zoom in, close enough to the gene-level so as to offer more robust and explanatory methodologies, but high enough to utilize information previously inaccessible to traditional approaches.

1.3 Systems Pharmacology

Systems pharmacology is a highly interdisciplinary and translational field at the confluence of systems biology and pharmacology, and one of the main application areas for pathway analysis methods. It combines experimental and advanced computational approaches, with the overarching goal of providing a comprehensive characterization of drug-induced perturbations and their links to human phenotypes (Berger and Iyengar 2009; Zhao and Iyengar 2012; Jenkins and Ma'ayan 2013). Its recent popularity is rooted in the availability of high-throughput molecular sequencing and screening technologies and the widespread access to the resulting wealth of data. Pharmacologically oriented systems biology involves high-throughput omics technologies, including next-generation sequencing, transcriptomics, proteomics, and metabolomics to pinpoint factors implicated in differential drug response across multiple scales of biological organization, ranging from molecular to cellular, tissue, and organism level. Drug action and system's response to it are embedded in the molecular interactions between the drug and its targets, which are further interacting with, and regulating, other cellular components. Under these circumstances, pathway analysis and modeling methods, offer not only the framework for integrating these heterogeneous data, but also a scaffolding for rendering them amenable to further investigations which can uncover properties of systems' components and relationships between them.

Pathway analysis approaches in systems pharmacology can be categorized into three broad groups based on their application areas.

Identification of Novel Drug Targets

Approaches under this category aim at finding new putative targets for drugs by identifying the relevant features of an optimal target. This is achieved by a comprehensive characterization of drugs' mechanism of action through: (i) investigating the pathways perturbed by treatment, (ii) defining the characteristics of the interaction network surrounding the drug target, (iii) as well as determining the proximity within the network to pathways affected by disease. This extensive understanding of drugs mechanism of action allows also the design of drugs, or drug combinations, able to act on multiple targets and pathways, which is a more efficient approach in targeting complex diseases, and the goal of an emerging research area known as polypharmacology.

Drug Repositioning

Pathway-based methods in this category use the mechanistic knowledge about drug targets and aim to retrieve associations to phenotypes distinct from the original drug indication. This is achieved by monitoring the biological pathways perturbed in response to drug and those affected by disease. The identification of common pathways or sub-pathways usually imply potential for repositioning.

Drug Safety and Side Effect Prediction

The contribution of pathway analysis in this direction is towards modeling the off-target (i.e., not originally intended as target) perturbation of drugs and their downstream effect. Off targets are widely accepted as the main source for drug side effects and toxicity. Thus, the possibility of predicting off-target interactions and the pathways or sub-pathways responsible for the propagation of the resulting perturbations, hold the promise of efficient treatment and reduced costs.

References

Ackermann M, Korbinian S (2009) A general modular framework for gene set enrichment analysis. BMC Bioinform 10:47

Barabási A-L, Réka A (1999) Emergence of scaling in random networks. Science 286:509–512

Berger SI, Iyengar R (2009) Network analyses in systems pharmacology. Bioinformatics 25:2466–2472

Chen X, Xu J, Huang B et al (2011) A sub-pathway-based approach for identifying drug response principal network. Bioinformatics 27:649–654

Cohen R, Havlin S (2003) Scale-free networks are ultra-small. Phys Rev Lett 90:058701

DiMasi JA, Hansen RW, Grabowski HG (2003) The price of innovation: new estimates of drug development costs. J Health Econ 22:151–185

DiMasi JA, Grabowski HG, Hansen RW (2016) Innovation in the pharmaceutical industry: new estimates of R&D costs. J Health Econ 47:20–33

García-Campos MA, Espinal-Enríquez J, Hernandez-Lemus E (2015) Pathway analysis: state of the art. Front Physiol 6:383

Goeman JJ, Bühlmann P (2007) Analyzing gene expression data in terms of gene sets: methodological issues. Bioinformatics 23:980–987

Gustafsson M, Nestor CE, Zhang H et al (2014) Modules, networks and systems medicine for understanding disease and aiding diagnosis. Genome Med 6:82

Hidalgo CA, Blumm N, Barabasi A-L, Christakis NA (2009) A dynamic network approach for the study of human phenotypes. PLoS Comput Biol 5:e1000353. doi:10.1371/journal.pcbi.1000353

Hung, JH, Yang TH, Hu Z et al (2011) Gene set enrichment analysis: performance evaluation and usage guidelines. Briefings in Bioinformatics bbr049

Ideker T, Galitski T, Hood L (2001) A new approach to decoding life: systems biology. Annu Rev Genomics Hum Genet 2:343

Jenkins SL, Ma'ayan A (2013) Systems pharmacology meets predictive, preventive, personalized and participatory medicine. Pharmacogenomics 14:119–122

Khatri P, Sirota M, Butte AJ (2012) Ten years of pathway analysis: current approaches and outstanding challenges. PLoS Comput Biol 8:e1002375

Meyers RA (2009) Encyclopedia of complexity and systems science. Springer, pp 719–741

Milgram S (1967) The small world problem. Psychol Today 2:60

Milo R, Shen-Orr S, Itzkovitz S, Kashtan N, Chklovskii D, Alon U (2002) Network motifs: simple building blocks of complex networks. Science 298:824–827

Petretto E, Liu ET, Aitman TJ (2007) A gene harvest revealing the archeology and complexity of human disease. Nat Genet 39:1299–1301

Wetterstrand KA (2016) DNA sequencing costs: data from the NHGRI genome sequencing program (GSP). Available at: www.genome.gov/sequencingcosts. Accessed 10 Dec 2016

Zhao S, Iyengar R (2012) Systems pharmacology: network analysis to identify multiscale mechanisms of drug action. Annu Rev Pharmacol Toxicol 52:505

Chapter 2
Networks and Pathways in Systems Pharmacology

Abstract This chapter presents an extensive overview of aspects involved in the thriving field of systems pharmacology. The three main directions along which network- and pathway-based analysis methods can contribute in systems pharmacology are spotlighted. Current approaches for the characterization of drugs mechanism of action, including the elucidation of mechanisms through which disease phenotypes dysregulate biological processes are first discussed. Subsequently, the latest research work done in systems pharmacology and polypharmacology toward the identification of novel drug targets, as well as in optimizing drug combinations for more efficient therapies, is surveyed. Within this context, the benefits of integrating evidence from multiple biological scales are examined, and the most popular databases used to store various biological data are provided. Drug repositioning is another direction along which pathway analysis is bound to bring significant contributions. An overview of drug repositioning approaches based on molecular and phenotypic profiles is presented. Subsequently, the main aspects involved in systems pharmacology applications for in silico drug side effect modeling and prediction are reviewed. Finally, current challenges and future considerations for pathway analysis and systems pharmacology are discussed.

Keywords Pathway analysis · Sub-pathways · Systems pharmacology · Drug mechanism of action · Polypharmacology · Drug repositioning · Drug targets · Drug safety · Drug side effect modeling · Heterogeneous data integration · Databases

2.1 Introduction

In the context of continuously surging drug development and healthcare costs, with the cost of developing a new drug being recently estimated at $2.8 billion, a more than 145% increase within the past decade only (DiMasi et al. 2016), and with global annual spending on prescription medication forecasted to reach $1.8 trillion, it becomes increasingly clear that the conventional reductionist *one drug*, *one target*, *one disease* paradigm which has been traditionally driving pharmacology

© The Author(s) 2017
A. Bezerianos et al., *Computational Methods for Processing and Analysis of Biological Pathways*, SpringerBriefs in Computer Science,
DOI 10.1007/978-3-319-53868-6_2

needs radical rethinking. Even with the significant raise in R&D expenditure by big pharma companies, the mean time between synthesis to approval surpasses 120 months, while the number of newly approved molecular compounds annually is ~ 20–30, not significantly different from what it was half a century ago (DiMasi et al. 2016; Csermely et al. 2013). Systems pharmacology brings the promise of revolutionizing the drug discovery process, while at the same time catalyzing the translation of pharmacogenomics applications to clinical environment, which has been lagging behind despite the recent wave of groundbreaking research on genomics implications in disease.

In this context, methods for modeling and analysis of molecular interaction networks, which have recently found extensive application in systems biology, are able to provide a theoretical platform for systems pharmacology. Studies on gene regulatory networks, protein–protein interaction networks, metabolic networks and other types of molecular interaction networks, provided significant insight into cellular organization and behavior, and shed light on specific biological processes, as well as disease processes and pathophysiology (Rual et al. 2005; Jeong et al. 2000; Ideker et al. 2002; Maraziotis et al. 2006, 2007; Bezerianos and Maraziotis 2008; Glaab et al. 2010). Consequently, based on this new network-based paradigm, new areas of translational research have emerged, and new terms have been coined, such as network physiology, network medicine, and network pharmacology (Barabasi et al. 2011; Hopkins 2008; Bashan et al. 2012).

Analysis of molecular interaction networks in systems pharmacology holds the promise of contributing along three main directions (Fig. 2.1):

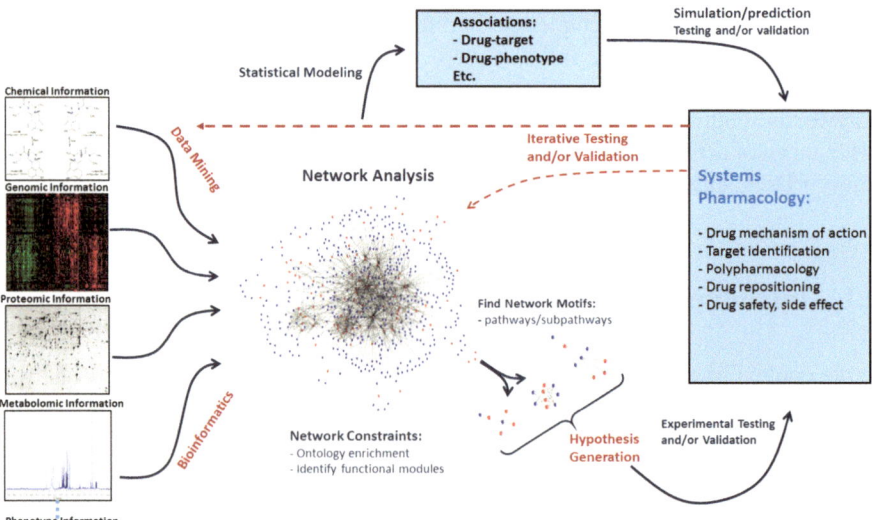

Fig. 2.1 Overview of network-based analysis in systems pharmacology [adapted from Arell and Terzic (2010)]

(i) Allowing for the identification of new putative drug targets relevant to specific diseases, through a better characterization of what makes an optimal target. In this context, pathway-based analysis allows a more mechanistic characterization of drugs *mechanism of action*, including the characterization of response to treatment, challenging the traditional way drug action was viewed: act on a specific target and observe the modulating effects downstream of that target. Current view is that a drug hits several targets (including off-targets) co-existing in a complex interacting network which is perturbed by disease, and the therapeutic effect of the drug aims to re-establish homeostasis (Berger and Iyengar 2009; Xie et al. 2009; Arell and Terzic 2010; Woo et al. 2015). Often times such approaches combine pathway and network analyses with pharmacokinetic and pharmacodynamic models to incorporate data from multiple biological scales, striving to build advanced quantitative and predictive models of therapeutic efficacy. As a corollary to achieving (i) follows the improved ability to predict effective drug combinations and the possibility to investigate mechanisms underlying drug resistance (Boran and Iyengar 2010; Zhao et al. 2013; Reddy and Zhang 2013; Lazar et al. 2014; Hwang et al. 2016).

(ii) *Drug repositioning* or *drug repurposing* is another direction in which systems pharmacology is making significant impact. Motivated by the success stories of several drugs with different initial indications, such as sildenafil (initially developed to treat hypertension and angina pectoris, eventually used to treat erectile dysfunction after clinical trial observations), or monoclonal antibody bevacizumab (originally developed to treat colon cancer and non-small cell lung cancer, currently used in treatment of macular degeneration disease), drug repositioning significantly shortens the path for approval of normal drugs and reduces the R&D expenditure (Van Eichborn et al. 2011; Wu et al. 2013; Pan et al. 2014; Li et al. 2016). Almost 20% of new drugs introduced to market in 2013 were actually new indications for existing drugs (Li et al. 2016). Originally based on serendipitous clinical observations, drug repositioning is picking up significant interest recently due to the increased understanding of the underlying molecular processes, drugs mechanisms of action, as well as the availability of advanced computational models for network and pathway-based analysis.

(iii) Another direction in which significant research efforts in systems pharmacology are focusing is that of *drug safety* and prediction of *drug toxicity* and *side effects*. Drug safety is a major source of drug attrition and of vital interest for pharmaceutical companies in their efforts to reduce drug development cost, while increasing efficiency (Hutchinson and Kirk 2011; Waring et al. 2015). Recent high profile failures during clinical trials or even for marketed drugs underline the fact that even efficacious drugs may cause severe side effects with dangerous consequences. Some examples include the cases of rosiglitazone, an antidiabetic drug which was later found to induce significant risk for myocardial infarction, rofecoxib, a pain relief drug recalled from the market after increased risk of stroke was reported, and the BIA-10-2474 (a molecule developed for a range of diseases) clinical trial death cases in 2016 (Graham et al. 2005; Nissen and Wolski 2007; Esserink 2016). It is therefore of paramount importance that molecular mechanisms of drug toxicity are comprehensively evaluated and used for hypothesis generation and testing, having as goal the development of in silico models for prediction of side effects. Systems pharmacology provides the framework for augmenting traditional pharmacokinetic and pharmacodynamics models while studying most common scenarios of drug toxicity from a pathway-based perspective: (a) off-target perturbations generating side effects unrelated to on-target effects, (b) side effects caused by pathways downstream of the intended on-target and (c) unrelated pathways generating side effects due to cross-talk with pathways downstream of intended target (Boran and Iyengar 2010; Wallach et al. 2010; Kuang et al. 2014; Lorberbaum et al. 2015; Cao et al. 2015; Trame et al. 2016; Schotland et al. 2016).

A concept with significant overlap to systems pharmacology, in both that it integrates systems biology with drug discovery and in its application areas, is *polypharmacology*. Polypharmacology includes studying the modulation of multiple targets by single drugs, as well as modulation of different targets by multiple drugs, primarily focusing on therapeutic interventions in complex diseases with the goal of identifying less toxic and more effective approaches (Boran and Iyengar 2010; Reddy and Zhang 2013; Anighoro et al. 2014). Another discipline that naturally converges to the more inclusive field of systems pharmacology is *pharmacogenomics*. Pharmacogenomics is defined by its search for variation in the human genome that explains inter-individual drug response variability (Antman et al. 2012). Currently in its incipient stage, with few genotype-drug response associations identified and finding their way into clinical practice by means of biomarkers present on drug labeling (FDA: Table of Pharmacogenomic Biomarkers in Drug Labeling 2016), translation of pharmacogenomic associations into clinical practice is still slowed by inconsistent findings and below par predictive power. Since these limitations are largely due to the complex interactions between drug-specific molecular response and environmental factors, systems pharmacology holds the promise to facilitate pharmacogenomics in unraveling the mechanisms behind the drug response variability. Rather than just identify mutations associated to diseases (e.g., genome-wide association studies), or perform statistical correlation type analysis between genetic signatures and patient phenotype, network- and pathway-based approaches of systems pharmacology allow integration of additional information for a better understanding of the bases of inter-individual variation, and in conjunction with pharmacogenomics, eventually lead toward the overarching goal of precision medicine (Turner et al. 2015).

The rest of this chapter is structured as follows: Sect. 2.2 describes current approaches in network and pathway-based characterization of drugs mechanism of action, Sect. 2.3 presents latest research work done in systems pharmacology and polypharmacology toward the identification of new drug targets, Sect. 2.4 provides an overview of systems pharmacology approaches in drug repositioning, Sect. 2.5

presents systems pharmacology applications for in silico drug side effect modeling and prediction. Final section presents current challenges and future considerations for pathway analysis and systems pharmacology.

2.2 Network- and Pathway/Sub-pathway-Based Characterization of Drugs Mechanism of Action

Initial efforts deviating from the traditional *one drug-one target-one disease* paradigm, and the related search for highly selective ligands that dominated the past decades, were triggered by the recognition that pharmacological compounds modulate the activity of targets in complex networks of deregulations underlying disease phenotypes (Gardner et al. 2003; Ambesi-Impiombato and Di Bernardo 2005; Hopkins 2008; Turner et al. 2015). These observations and the ensuing endeavor for investigating the compounds mechanisms of action (MoA) were only possible with the advent of high throughput technologies which started generating wreaths of data and with the concomitant rise of the new field of systems biology (Ideker et al. 2001).

The elucidation of mechanisms by which drug compounds affect the deregulated interactions in disease phenotypes is bound to become an essential part of the modern drug discovery process. With this comes an increased need for computational methods to mine large datasets and assist in providing initial hypotheses for further in vitro and in vivo validation studies. About a decade ago, data resources originating from genome-wide transcriptional profiles and containing drug response phenotypes, such as the Connectivity Map (CMap—which contains more than 7000 gene expression profiles obtained in response to treatment with 1309 drug and drug-like small molecules) became available, followed in recent years by similar databases, such as the Library of Integrated Network-based Cellular Signatures project (LINCS) (Lamb et al. 2006; Wang et al. 2016). The use of gene expression data (transcriptional mRNA profiles, initially obtained from microarray experiments, more recently from RNAseq experiments) in investigating drugs' MoA has become norm, as this type of data allows genome-wide investigation of drug response' correlation with disease phenotype. Early work successfully characterized compounds perturbation mechanisms by searching for commonalities in the phenotypic responses based on the simple hypothesis that, if two drugs induce similar transcriptional responses they potentially share a common MoA and a similar therapeutic application, even if they act on different cellular target (Kibble et al. 2016). This idea was adapted from early investigations in genomic data analysis in which it was observed that genes with similar expression profiles are more likely to be involved in common biological processes. Transcriptional response profiles were initially compared using various methods similar to the Gene Set Enrichment Analysis (GSEA), based on the Kolmogorov–Smirnov statistic (Subramanian et al. 2005). Briefly, query signature profiles' similarity to the reference expression profiles in the CMap database is assessed. Query profiles are usually sets of genes differentially expressed between disease and normal conditions, or sets of up- and/or down-regulated genes. In parallel, genes on the reference CMap arrays (each one corresponding to experiments in which cells are perturbed using a specific drug) are rank-ordered according to their differential expression relative to control. Subsequently, the query signature is compared to every rank-ordered gene list and it is determined whether up-regulated genes in the query tend to be located near the top of the list and down-regulated genes are found toward the bottom of the reference ranked list, or vice versa. The former case denotes a 'positive connectivity' and the latter a 'negative connectivity' between the query and the respective perturbation instance (array containing the cells gene expressions in response to the drug treatment). Connectivity scores are then computed and used to rank all instances in the database according to their correlation to the query signature. This approach was used by Lamb et al. (2006) to elucidate the MoA of uncharacterized drug compounds, such as gedunin. The mechanism through which gedunin is capable of abrogating the expression of androgen receptor (AR) activation in prostate cancer was determined by finding high connectivity scores of a gedunin signature with multiple instances of three heat shock protein 90 inhibitors (HSP90): geldanamycin, 17-allylamino-geldanamycin, and 17-dimethylamino-geldanamycin (Lamb et al. 2006). It was therefore inferred that gedunin might impinge upon the HSP90 pathway, hypothesis which was subsequently validated experimentally. This hypothesis would not have been warranted by solely studying compounds structures, as gedunin is structurally dissimilar to known HSP90 inhibitors.

Various other approaches based on ranked lists of differentially expressed genes, have been used, such as the MANTRA method (Iorio et al. 2010), which adopts a rank-aggregation procedure to dilute cell-line-specific effects in transcription, as well as experimental batch effects, or different drug concentrations in different treatment instances. Iorio et al. (2010) defined pairwise distances between compounds using 'enrichment scores' based on the distribution of optimal gene signatures of each compound (extracted as top and bottom 250 genes in their corresponding ranked lists) within the ranked gene list of the other compound of the pair and vice versa. These distances were used to build a drug network in which nodes correspond to compounds and connecting edges reflect the estimated distances between the compound pairs. This network was subsequently mined via network clustering to identify communities (or modules) of closely interconnected compounds. The retrieved drug modules were found to be highly enriched with common biological pathways and characterized by similar MoAs. The authors have then proceeded to predict MoA for anticancer drugs with profiles not present in the reference CMap database, by estimating the distance of their transcriptional profiles to the drug network modules. Following this framework, PHA-690509, PHA-793887, and PHA-848125 were correctly classified as CDK inhibitors, distinct from the other kinase inhibitors in the CMap database, and were also predicted to have highly similar MoA to Topoisomerase inhibitors. The original method in (Iorio et al. 2010) was recently extended to filter out spurious effects of compounds' nonspecific secondary effects on transcriptional profiles. To this goal, they use an iterative supervised approach to refine the original drug network module of a compound of interest while deriving a transcriptional signature representative of the primary MoA (Iorio et al. 2015).

Some studies have argued that methods based solely on differentially expressed sets of genes (i.e., transcriptional profiles) may miss essential knowledge on regulatory influence among genes and their products. Consequently, methods such as the mode-of-action by network identification (MNI), which incorporates differential expression of genes with regulatory information encoded in gene networks structures, have been proposed (Xing and Gardner 2006). In MNI systems of linear differential equations are used initially to build the gene network model and the subsequent inference of network parameters is done based on transcriptional profiles. Once the canonical gene network is created, it is used to filter test transcriptional profiles from drug treatment experiments in order to distinguish genes that are mediators of treatment response from the other genes which exhibit expression changes. This is achieved by searching for genes with changes in their transcriptional profiles that are not in accordance with the canonical gene network, under the assumption that such genes are perturbed by the drug treatment. Significance of the perturbation on these putative molecular targets is quantified using a z-score scheme. MNI was utilized to identify molecular targets of antifungal compounds based on genome-wide transcriptional profiles in yeast.

Recently, it was proposed that data from additional sources, such as signaling and metabolic pathway databases, protein structure databases, compound structure and drug target databases, as well as DNA sequence or functional non-coding RNA, may be incorporated in the analysis. This integrative approach has the potential to enrich the computational model, by making it more biologically plausible, and enhance its predictive power (see Fig. 2.1). Table 2.1 presents some of the most commonly used databases containing data and annotations involved in MoA identification and generally in drug discovery. Within this context, Iskar et al. (2013) used bi-clustering to identify drug-induced transcriptional modules from human and rat transcriptional profiles databases [CMap and DrugMatrix (Ganter et al. 2005)]. The modules conserved across organisms were checked for functional coherence at protein level using information from the STRING database (Szklarczyk et al. 2014) and then connected into a module network. The module network was extensively characterized by annotation with relevant pathways and functional information from KEGG (Kanehisa et al. 2015), BioCarta (Nishimura 2001) and the Gene Ontology (Gene Ontology Consortium 2013) databases, as well as drug structure, target, and side effect information from STITCH and SIDER databases (Szklarczyk et al. 2015; Kuhn et al. 2015). The integrative model thus defined allowed the authors to discover novel MoAs for six drugs, four with cell-line-specific mechanism and two with mechanisms conserved in all modules, using module-based statistical tests and overrepresentation analysis. Specifically, zaprinast, was suggested to be a novel modulator of the PPARγ receptor in the PC3 cell line, the main target of antidiabetic drugs, a hypothesis subsequently validated with target binding assays experiments. Similarly, nitrendipine was found to be a modulator of estrogen receptor in MCF7 cells, hexetidine and (+)-chelidonine were

Table 2.1 Public databases commonly used in systems pharmacology

Database	Web address	Notes
DrugBank	http://www.drugbank.ca/	Richly annotated bioinformatics and cheminformatics resource that combines detailed drug data (e.g., chemical, pharmacological, and pharmaceutical) with comprehensive target information (e.g., sequence, structure and pathway)
TTD	http://bidd.nus.edu.sg/group/cjttd/TTD_HOME.asp	Therapeutic target database (TTD) provides the information about known and explored therapeutic protein and nucleic acid targets, the targeted diseases, pathway information and corresponding drugs directed at each of these targets
SuperTarget	http://insilico.charite.de/supertarget/	SuperTarget is an extensive database for analyzing drug–target interactions. This database allows querying by drugs, targets, drug–target-related pathways, drug–target-related ontologies
STITCH	http://stitch.embl.de/	Database of known and predicted chemical–protein interactions, which integrates the evidence derived from experiments, other databases, and literature
BindingDB	http://www.bindingdb.org/bind/index.jsp	The BindingDB is a binding database of experimentally determined protein–ligand binding affinities among small molecule ligands and protein targets
SIDER	http://sideeffects.embl.de/	Public and computer-readable database that contains information on marketed drugs and their recorded side effects (i.e., adverse drug reactions), including side-effect frequency, drug and side effect classifications as well as drug–target associations
OFFSIDES/TWOSIDES	http://tatonettilab.org/resources/tatonetti-stm.html	Offsides database is a resource of off-label side effects not listed on the FDA's official drug label. Twosides databases is a resource of polypharmacy side effects for pairs of drugs
FAERS	http://www.fda.gov/Drugs/GuidanceComplianceRegulatoryInformation/Surveillance/AdverseDrugEffects/ucm082193.htm	FDA database that contains the information obtained from adverse event and medication error reports submitted to FDA on side effect keywords (adverse event keywords) for drugs

(continued)

Table 2.1 (continued)

Database	Web address	Notes
MATADOR	http://matador.embl.de/	Database resource for protein–chemical interactions, including multiple direct and indirect modes of drug–target interactions. The manually annotated list of direct (binding) and indirect interactions between proteins and chemicals was assembled by automated text mining followed by manual collection
ChEMBL	https://www.ebi.ac.uk/chembl/	Literature-derived database of molecule structures and molecule–protein interactions. This includes a catalog of approved drugs
ChemBank	http://chembank.broadinstitute.org/	Database containing collections of data derived from small molecules and small molecule screens, and resources for studying their properties
PubChem	https://pubchem.ncbi.nlm.nih.gov/	Public repository of biological assay results for hundreds of thousands of molecules
canSAR	https://cansar.icr.ac.uk	Cancer-related resource on biological annotations, chemical screening, RNA interference screening, expression, amplification and 3D structural data
Gene Ontology	http://geneontology.org/	Bioinformatics initiative to unify the representation of gene and gen product attributes across all species. Aims to maintain and develop its controlled vocabulary of gene and gene product attributes, annotate genes and gene products
STRING	http://string-db.org/	Database of known and predicted protein–protein interactions. The interactions include direct (physical) and indirect (functional) associations; they stem from computational prediction, from knowledge transfer between organisms, and from interactions aggregated from other (primary) databases
CMap	http://portals.broadinstitute.org/cmap/	Gene expression profiles of 1309 drugs across human primary cell lines
Gene Expression Omnibus (GEO)	https://www.ncbi.nlm.nih.gov/geo/	NIH public repository of functional genomics data

(continued)

Table 2.1 (continued)

Database	Web address	Notes
LINCS	http://www.lincsproject.org/	Large scale NIH repository containing gene expression profiles of drugs and genetic perturbagens across numerous cell lines
DrugMatrix	https://ntp.niehs.nih.gov/drugmatrix/index.html	Molecular toxicology reference database; is populated with the comprehensive results of thousands of highly controlled and standardized toxicological experiments in which rats or primary rat hepatocytes were systematically treated with therapeutic, industrial, and environmental chemicals at both non-toxic and toxic doses
PharmGKB	https://www.pharmgkb.org/	An interactive tool for researchers investigating how genetic variation effects drug response. Contains corresponding drug–gene relations
KEGG	www.genome.jp/kegg/	Resource for understanding high-level functions and utilities of the biological system from molecular-level information. Contains signaling pathways, drug, structure and chemical information
IntAct	http://www.ebi.ac.uk/intact/main.xhtml	Database of evidence for molecular interactions, and maintains the reference resource for macromolecular complexes
HINT	http://hint.yulab.org/	Database of high-quality protein–protein interactions in different organisms. These have been compiled from different sources and then filtered both systematically and manually to remove erroneous and low-quality interactions. HINT can be used for individual queries as well as for batch downloads
HPRD	http://www.hprd.org/	Database with information on human protein functions including protein–protein interactions, post-translational modifications, enzyme-substrate relationships and disease associations
BIOGRID	https://thebiogrid.org/	Public database that archives and disseminates genetic and protein interaction data from model organisms and humans

(continued)

Table 2.1 (continued)

Database	Web address	Notes
DIP	http://dip.doe-mbi.ucla.edu/dip/Main.cgi	Database which catalogs experimentally determined interactions between proteins. It combines information from a variety of sources to create a single, consistent set of protein–protein interactions. The data stored within DIP have been curated, both manually, by expert curators and automatically, using computational approaches
PathwayCommons	http://www.pathwaycommons.org/	Database of biological pathways and interactions from multiple organisms
BioCarta-PID	http://cgap.nci.nih.gov/Pathways/BioCarta_Pathways	NIH NCI resource on pathways and pathways interaction
Wiki Pathways	http://www.wikipathways.org/index.php/WikiPathways	Open, collaborative platform dedicated to the curation of biological pathways. WikiPathways thus presents a new model for pathway databases that enhances and complements ongoing efforts, such as KEGG, Reactome and Pathway Commons
Reactome	http://www.reactome.org/	Open-source, curated and peer reviewed pathway database
TarBase	http://diana.imis.athena-innovation.gr/DianaTools/index.php?r=tarbase/index	Provides high-quality manually curated experimentally validated miRNA:gene interactions, enhanced with detailed meta-data
miRGen	http://carolina.imis.athena-innovation.gr/diana_tools/web/index.php?r=mirgenv3%2Findex	miRGen is an integrated database of miRNA gene transcripts, transcription factor binding sites, miRNA expression profiles and single nucleotide polymorphisms associated with miRNAs
HMDB	http://www.hmdb.ca/	Database containing small molecule metabolites found in the human body
BRENDA	http://brenda-enzymes.info/	Enzymes database
OMIM	www.omim.org/	Online catalog of human genes and genetic disorders containing gene–phenotype associations
ConsensusPathDB	http://consensuspathdb.org/	Integrates interaction networks in Homo sapiens including binary and complex protein–protein, genetic, metabolic, signaling, gene regulatory and drug–target interactions, as well as biochemical pathways

found and experimentally confirmed to have adrenergic activity. Additionally, the same study identified novel functions for 10 previously poorly characterized genes as modulators of cholesterol homeostasis, based on their strong connections within the transcriptional modules enriched for cholesterol biosynthesis pathways.

Using an approach that attempts to both capture regulatory information encoded in the interaction network and integrate various levels of information (transcriptional, signaling, and protein-level interactions), Woo et al. (2015), extend the cell type-specific approach to a tissue-specific one. To this goal they build lymphoma-specific regulatory networks based on transcriptional profiles on in vivo and in vitro drug perturbations. Their approach incorporated translational level information (protein–protein interaction data) and protein–DNA interaction data to create the contextualized regulatory network. MoAs are characterized by modeling and quantifying compounds' dysregulation of network neighborhoods using a probabilistic framework based on Gaussian kernel smoothing. The approach allows the authors to mechanistically elucidate MoAs, while accounting for differential expression of associated nodes (genes or proteins) from a network-based perspective, rather than a purely statistical one. Their study highlighted key differences in topoisomerase (TOP) inhibitor compounds doxorubicin, camptothecin, and etoposidine, which all have previously known significant common footprint. The identified specific effectors were validated experimentally, confirming the approach's high specificity. The same method was used to identify novel compound effectors and modulators for vincristine (a microtubule formation inhibitor in mitotic spindle), mitomycin C, and altretamine (antineoplastic drugs).

A relatively recent trend in pathway analysis, as highlighted in the previous chapters, is that of sub-pathway-based approaches. Investigating sub-pathways may be more relevant in interpreting the biological processes, since it is known that, frequently, only some regions of pathways are dysregulated by disease, or involved in drug related perturbations. Within this context, Chen et al. (2011) have devised a method to identify sub-pathways involved in dexamethasone (DEX) response in human prostate cancer cell lines. Their approach relied on parsing sub-pathways from the KEGG Pathway database in an exhaustive manner. Sub-pathways were defined as individual paths from start points to end points in a pathway map. Such an approach is biologically relevant, as pathway maps in KEGG database are linear sequences connecting biologically meaningful start nodes (which are commonly membrane receptors or their ligands) to end points which are commonly transcription factors or their targets. The resulting sub-pathways were overlaid with transcriptional profile data of a subset of CMap (instances of DEX treated cells). In order to identify sub-pathways significant for DEX-response, a two-stage approach was followed, by defining aggregate distances between sub-pathway states pre- and post-treatment in terms of their contained genes expression levels, and subsequently identifying through statistical analysis key subsets of genes most perturbed by drug, and therefore deemed top contributors to the sub-pathway state differentiation. Based on this, authors were able to assert that the decrease of VEGFR and EGFR stabilization in order to suppress angiogenesis is a hallmark of DEX-response.

Pritchard et al. (2013) proposed an innovative analysis pipeline based on supervised and unsupervised machine learning methods with the goal of achieving both statistical and biological generalization (predictive accuracy), and at the same time ensure the ability of their framework to recognize novel MoAs for drugs. To this goal they define the drug MoAs in terms of subnetworks consisting of drug nodes and edges representing weighted connections between nodes. The weighted connections correspond to distances in the molecular signature space. Initial subnetwork membership is based on biochemical and genetic evidence encompassing three different types of data: mRNA, chemical interaction and RNAi and each subnetwork corresponds to a drug MoA. The training set corresponds to subnetworks of known drug MoAs. Given a test set of uncharacterized drugs, predictions are made based on a k-nearest neighbors method, and putative MoAs are obtained based on sets of representative features corresponding to subnetworks in the training set. A prediction may interpolate within an existing subnetwork or extrapolate to form a new expanded subnetwork. Detection of new MoAs is warranted when a too large expansion of subnetworks is needed. By using a consensus approach the method identifies new clusters within the training set drugs, based on their molecular features. Subsequently, unsupervised learning (hierarchical clustering) is utilized to identify optimal topological thresholds for the connecting edges within the newly derived subnetworks. The procedure enables the detection of more than mere combinations of existing subnetwork motifs, thus permitting the extension to MoAs underlying entirely distinct biology. Using this subnetwork-based signature, authors confirmed MoA subnetworks for HSP90 and EGFR inhibitors suggested in previous studies. Additionally, they were able to successfully confirm and expand MoA classes including erastin (a Bax/Bak independent death inducing compound), mitochondrial disruptors azide and valinomycin and predict mitoxantrone as a topoisomerase II poison.

A more recent approach which exploits relationships shared between drugs within a network context is presented in (Napolitano et al. 2016). Namely, the method extends the GSEA framework to define enrichment scores for pathways across sets of drugs. It eventually produces ranked lists of drugs highlighting the potential for dysregulation induced in specific pathways by specific sets of drugs. The method, termed drug-set enrichment analysis (DSEA) incorporates pathway information from various related databases to essentially produce a pathway-based connectivity map. This enabled the authors to formulate hypotheses on the MoAs shared by drugs. Thus, DSEA was utilized to identify shared pathways by sets of drugs in five distinct pharmacological classes with known MoA and results were validated by means of gold standard sets of target genes for each class retrieved from molecular databases. Additionally, the method was able to infer a putative MoA for a set of drugs with mild corrective activity in cystic fibrosis, a disorder for which no therapeutic treatment is currently available. The approach has the potential for aiding in the characterization of novel drugs with unknown MoA by simply incorporating related transcriptional profiles into the pipeline.

2.3 Identification of New Drug Targets and Polypharmacology Applications

Systems pharmacology approaches for inferring compound MoA have evolved in the past decade from methods based purely on ranked lists of genes and their transcriptional response to treatment, to gradually incorporate elaborate network and pathway context, as well as various other sources of biological information. The rising interest in understanding compound MoA was accompanied by a simultaneous strive for identifying novel therapeutic targets based on network analysis methods, within the greater context of optimizing the drug discovery process. Computational methods based on network analysis can be used to model the systemic milieu in which putative therapeutic targets are located and consequently identify targets which increase therapeutic efficacy and reduce adverse effects. In order to achieve this goal, the complex relationships between the chemical and genomic factors influencing the interaction between drugs and their targets must be appropriately accounted for.

From this perspective, the concept of similarity among various biological and nonbiological entities (such as compound chemical structure, protein sequence, phenotypic profiles, etc.) is paramount. Similarity is at the base of two important hypotheses in modern drug discovery, in the sense that chemically and pharmacologically similar drugs are targeting similar target proteins (Chen et al. 2012), and that molecularly and clinically related drugs and diseases are likely to share similar phenotypes (Vogt et al. 2014). Additionally, in the context of systems pharmacology, multifaceted similarity metrics can be used to facilitate the integration of heterogeneous data. As in the case of approaches used in MoA identification, networks built for the identification of novel targets have edges representing protein–protein interactions and transcriptional regulation but may also encode drug–target or drug–drug interactions. Commonly, edges are defined based on therapeutic or chemical similarities between two nodes, similarities between proteins sharing associations with diseases, or similarities of diseases based on the shared number of genes/proteins (Zhao and Iyengar 2012). This wide range of possible definitions for network edges, and their underlying similarity metrics, enable networks to model multiple interaction scales, transcending from atomic and molecular level to the phenotype level of drug–target interactions.

2.3.1 Target Characterization and Identification Using Network Properties of Drug Targets

Since an important part in the process of identification of novel drug targets is the understanding of how signal flow is achieved within molecular pathways, a significant share of research work in this area has been dedicated to studying network topology-based relationships and identification of target-related motifs. Additionally, concepts such as network paths are important for establishing relations between nodes and network topologies, and formulate biologically relevant constraints in modeling drug perturbation (e.g., start nodes on a path must be receptors, intermediate nodes be specific types of intracellular proteins and end nodes must be transcriptional factors). Such methods rely on interaction networks built from protein–protein data on which drug-related data is overlaid, or on bipartite or multipartite networks used to model drug–target and drug–drug interactions (Yildirim et al. 2007; Yamanishi et al. 2008; Li et al. 2015). Early work focused on formulating network topology criteria which define existing drug targets and, based on these criteria, elaborate methods that would allow the identification of novel targets from the network (Yildirim et al. 2007; Ma'ayan et al. 2007; Yamanishi et al. 2008; Hwang et al. 2008; Nacher and Schwarz 2008; Berger and Iyengar 2009). Yildirim et al. (2007) used a bipartite network based on two projections: in the first, nodes denote drugs which have connecting edges if they share a common target, while in the second projection nodes denote protein targets which are connected if they share a common drug. The analyses of these networks revealed that drug targets tend to have a higher degree (number of connecting edges) than other nodes, and therefore are implicated in more cellular interactions. Additionally, they observed that most new drugs are associated with previously targeted network neighborhoods. Ma'ayan et al. (2007) used a bipartite network connecting drugs and drug targets, overlaid on protein–protein interaction data to show that drug target proteins are primarily located in the cellular membrane. Another important observation derived from the topology-based studies is that network centrality or node degree measures should not be the sole factors for the detection of new target proteins. Although such measures indicate essentiality of respective protein nodes, perturbation induced by drug treatment on the respective protein targets could induce significant undesired effects on the downstream cellular processes. Hwang et al. (2008) instead proposed targeting proteins which are bridging nodes with less regulatory effects on pathways (fewer interacting connections), but located in network positions where their disruptions would result in information flow prevention.

More recently, Mitsopoulos et al. (2015) identified sets of topological and community properties characterizing druggability of target protein nodes and neighborhoods and highlighted differences between cancer and non-cancer drugs. To this goal they used protein–protein interaction data enriched with drug-target information and built sets of predictors based on the network topology descriptors. Machine learning methods such as random forests, gradient boosted machines, and generalized linear models were then utilized to computationally validate their drug–target interaction predictions. In Li et al. (2015) authors define a computational framework based on the guilt by association principle and network topology features, which allows them to identify a large number of potential drug targets, among which some are associated with diseases such as the Torg-Winchester syndrome and rhabdomyosarcoma. Under the guilt by association assumption, a target protein and a drug are likely to interact if the majority of the protein's neighbors (which share direct interactions with the target protein) in the network can interact with the drug. The authors use a predictive model based on the random forest algorithm and feature sets consisting from node and edge weights in a bipartite network model (containing protein–protein, drug–target and drug–drug interactions).

2.3.2 Identification of Drug Targets Based on Integrative Network Approaches

The task of identifying drug targets from genome-wide data can be helped by the integration of additional data such as drug chemical structure, target protein sequences, known drug–target interactions, or information about drugs' side effects. As in the case of MoA characterization, the incorporation of such complementary data can help in adding more biologically plausible context to the models, reduce bias induced by incomplete information and enhance the search space for the computational algorithms deriving the predictive models.

Campillos et al. (2008) proposed a method incorporating information on drugs' side effects from drug package inserts into a drug–target network in order to define a phenotype-based similarity metric. The side effect similarity metric was combined with a 2D chemical similarity metric based on the Tanimoto coefficient into a probabilistic framework under which to infer the probability of two drugs interacting with the same target. The method was used to derive new targets for existing drugs, and the authors validated using in vitro assays 13 drug–target interactions predicted by their method. However, the main limitation of such an approach was that it could only be used on marketed drugs for which side effect information was available.

Based on the same experimentally validated assumption that similar drugs interact with similar target proteins, Chen et al. (2012) integrated a composed drug–drug similarity metric based on drug chemical structure similarity and targets known to be shared by pairs of drugs, a target–target similarity metric based on protein sequence similarity, and a known drug–target interaction network. The authors then implemented a random walk with restart on the resulting bipartite drug–target network to predict potential drug–target interactions. Thus, a target can be predicted even if the investigated drug has no known targets, based on similar drugs and their known targets. The random walk was implemented using transition matrices from target network to drug network and inter-transition matrices indicating the probability of walks from drug to drug (or target to target). Based on this, a probability of finding the walker at node i at step $t + 1$ based on the position at node j at step t can be determined iteratively. The approach was used to predict drug target interactions for four classes of datasets (enzymes, ion channels, G protein coupled receptors and nuclear receptors). Results were validated using gold standard datasets from public databases.

Cheng et al. (2012) combined three supervised inference models to predict drug–target interactions. Namely, the network-based inference (NBI) relying on drug–target bipartite network topological similarity was used in conjunction with a drug based similarity inference (DBSI), which relies on 2D chemical similarity between drugs and drug–target interaction information, and a target-based similarity inference (TBSI), relying on target sequence similarity and drug–target interaction information, to predict associations between drug–target pairs. DBSI and TBSI incorporated information from the chemical and genomics space, respectively, while NBI was based solely on network topology features. The authors highlighted the performance of NBI inference, superior to the other inference methods. The predicted targets were validated using in vitro binding assays. The approach indicated polypharmacological effects on five drugs (montelukast, diclofenac, simvastatin, ketoconazole, and itraconazole) and suggested repositioning potential of these drugs, which was further validated experimentally.

An interesting approach has been recently proposed by Isik et al. (2015), which investigated the transcriptome perturbations in conjunction with functional interaction network information to reveal effects induced by drugs binding to their targets. They derive a new measure for target prioritization, termed local radiality, which is able to identify more diverse targets, with fewer neighbors, and consequently, possibly fewer side effects. They validate the results based on ROC analysis using test datasets from other approaches.

A large number of other network-based and machine learning-based methods have been developed recently, most of them following broadly the same paradigm, as shown in Fig. 2.2: enrich existing networks of known drug–target interactions with information from chemical and/or genomics spaces and learn various

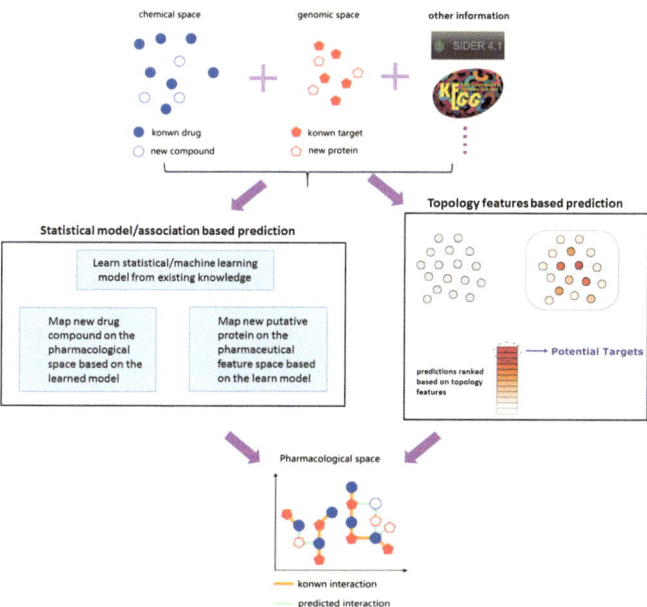

Fig. 2.2 Schematic of target identification approaches in systems pharmacology

supervised or semi-supervised models in order to predict novel interactions. For example Yamanishi et al. (2008) used a kernel regression method to learn chemical and genomic space models and demonstrate correlation with pharmacological space. Yuan et al. (2016) used a similarity approach based on ensemble learning methods to incorporate chemical and genomic space similarity as components into ensembles of learning to rank, while Yamanishi et al. (2014) created a web based engine (DINIES) using supervised learning and relying on similarity matrix kernels (learned from drug, side effects and protein domains) to predict interactions on test sets. Another recently developed web-based tool is TarPred (Liu et al. 2015), which besides predicting targets is also able to provide disease indications and predict side effects.

2.3.3   Network-Based Polypharmacology

It is often the case that methods developed for predicting new drug target inter-actions naturally yield combinations of potential targets (often protein complexes or whole sub-pathways), which naturally classifies them as polypharmacology approaches. Polypharmacology accounts for the important and increasingly accepted concepts that (i) complex diseases tend to be associated with multiple target proteins, and (ii) drugs commonly work by targeting several off-targets, besides the primary target (Xie et al. 2012). Accounting for the polypharmacology properties of drugs has the potential of increasing drug efficacy and overcoming drug resistance and toxicity, thus, the approaches capable of developing multiple target drugs, as well as research in drug combination based on network approaches have received increased attention recently.

An example is the method developed by Yang et al. (2008), which developed a computational framework for inferring multiple targets and suggest optimal com-binations of target intervention. Their method, named multiple target optimal intervention (MTOI), searches systematically for effective points of intervention in a disease-based network to restore it to a desired normal state. MTOI relies on a procedure for perturbing the disease network and optimize it toward the desired state based on a Monte Carlo simulated annealing optimization algorithm (MCSA). The disease network is defined as a collection of concentrations of proteins and/or metabolites, or other relevant temporal-based information. Such a network is usu-ally obtained from experimental data on patients or cells in abnormal/disease condition. The desired network is defined as the physiological steady state network. The information and related perturbations are modeled using differential equations and MCSA. The authors applied it to an inflammation based network, the arachi-donic acid metabolic network, and derived a combinatorial intervention based on anti-inflammatory drugs.

Other network based polypharmacology studies include that of Cheng et al. (2012b), which extended their previous method (Cheng et al. 2012a) and proposed two different weighted network-based inference methods using four similarity metrics for predicting multiple chemical–protein interaction. Under this framework they investigated the polypharmacology of five approved drugs: imatinib, dasatinib, sertindole, olanzapine, and ziprasidone. Zhao et al. (2013) used a composite network built from protein–protein interactions and gene regulatory databases onto which Gene Ontology and side effect information was overlaid. Drug–drug pairs, for which the addition of a pair member was reported to result in reduced side effects of the other drug, were exhaustively searched for. Random walk was then used to determine interaction subnetworks between drug pairs, in order to identify nodes that would be preferentially affected by specific interactome perturbations. Following this approach the authors were able to predict drugs which combined with rosiglitazone (an efficacious antidiabetic drug associated with increased myocardial infarction), would mitigate its myocardial infarction risk. Additionally, they predicted that the mitigating effect of exenatide in conjunction with rosiglitazone could occur through clotting regulation. Additional polypharmacology-related approaches are presented extensively in review studies (Reddy and Zhang 2013; Medina-Franco et al. 2013).

It must be noted at this point that usually there are significant overlapping areas between approaches attempting drug–target interaction prediction, polypharmacology-related methods and methods having as goal repositioning strategies for existing drugs. It is often the case that, due to the limited available resources on drugs, target-identification methods are restricted to predict alternate targets for drugs with already known targets, which is essentially a drug repositioning approach. This is the case with methods developed in Campillos et al. (2008) and Cheng et al. (2012a) described above. The same stands for studies investigating drugs MoA, which commonly have as byproduct multiple genes/proteins, often representing entire sub-pathways identified as target of a specific drug (Iskar et al. 2013; Chen et al. 2011), which could be seen as polypharmacology studies. In turn, the search for polypharmacological features naturally leads to new uses for combinations of known drugs, thus providing support in drug repositioning (Chen et al. 2015).

2.4 Network-Based Drug Repositioning

Drug repositioning research has gained significant momentum in recent years due to the pressing needs to reduce costs of developed drugs while increasing efficacy, but also due to large-scale funding programs launched by governmental organizations, such as the National Center for Advancing Translational Sciences and FDA in US, and the Medical Research Council in UK (Li et al. 2016). Drug repositioning is inherently linked to a better understanding of the molecular context underlying specific phenotypes and of the mechanisms of action of drugs, which are additional reasons for drug repositioning approaches to be flourishing with the advent of systems pharmacology.

A ubiquitous feature in network-based drug repositioning is the presence of a disease-related component, since finding associations between drugs and protein targets in a disease context, is the modus operandi in such studies (Wu et al. 2013). Therefore, three level drug–target–disease networks are a common occurrence in modern drug repositioning research. Effective network-based approaches typically aim at accurately modeling the cause-effect paradigm which is dominating the current view on disease etiology and drug mechanisms of action: disease originates from abnormalities of one or more (usually genetic) factors and the observed phenotypes are the effect of disease development. Similarly, drug action originates from the drug–target binding and the terminal effect of the drug intake are the drug indications and side effects, which can be seen as drug phenotypes. Along these cause-effect paths, molecular activities induced by drugs and diseases may be observed using high throughput transcriptional and proteomic data, which can be viewed as snapshots of the disease development stages or of drug activity, and consequently be used to model drug–disease associations (Li et al. 2016). From this perspective drug repositioning studies can be categorized as being molecular profile based or phenotype profile based.

2.4.1 Drug Repositioning Based on Molecular Profiles

Generally, drug repositioning approaches based on molecular profiles of drugs and/or diseases rely on the so-called reversed signature hypothesis: if the molecular profile of a drug is opposite to that of the disease, then the drug has the potential to be used in treating that disease (Wu et al. 2013; Li et al. 2016). Work in this area typically follows the now standard procedure in systems pharmacology: first construct a background interaction network from protein–protein interaction databases, pathway databases, protein–DNA interaction databases, and/or other interaction resource available. Then contextualize the initial network, for example by adding weights to the edges leveraging gene expression data from sources such as CMap, LINCS, or GEO, or enriched with data from various other sources (GO, KEGG, etc.). Subsequently, various computational models and algorithms can be used to extract parts of the contextualized networks (response subnetworks or sub-pathways) which maximize the biological relevance related to disease–drug associations (Fig. 2.3).

Following such an approach, Jin et al. (2012) created their interaction network from signaling pathways in PID and BioCarta databases, onto which they overlaid transcriptional data from CMap and subsequently searched for network motifs (sub-pathways) involved in response to cancer drug treatment. These sub-pathways are connecting the disease genes (retrieved from OMIM) to known signaling proteins. They used Bayesian factor regression to uncover such driver sub-pathways bridging drug targets to the disease response signatures. The driver sub-pathways and the drug's effects on them were found simultaneously. The effect of drugs on each sub-pathway was quantified and summarized into drug–disease signature

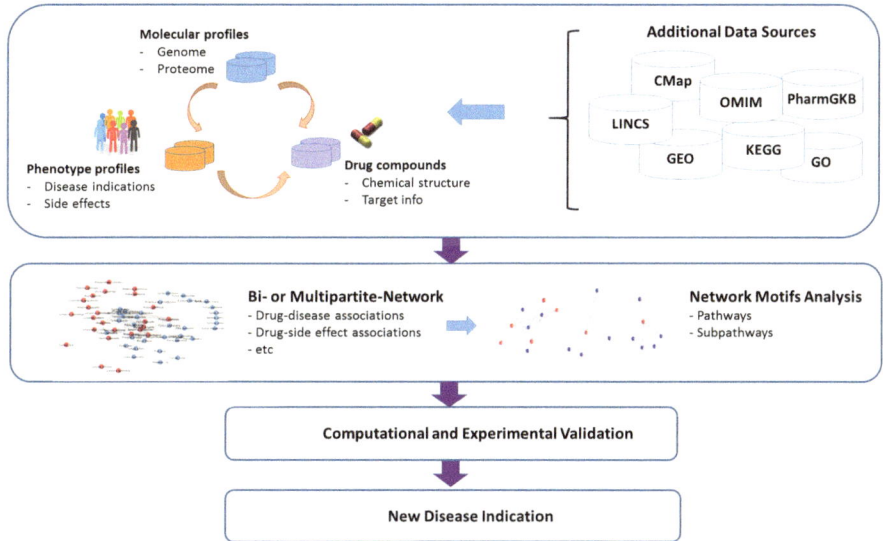

Fig. 2.3 Typical drug repositioning workflow

profiles. Then, ranked repositioning profiles for each of the drugs were created and repositioning potential derived accordingly, using support vector regression. Several high ranking drugs from their analysis were suggested for repositioning in cancer therapy based on the ability to enforce retinoblastoma-dependent repression of important E2F-dependent cell-cycle genes (Jin et al. 2012). Additionally, their method was able to accurately predict responses to more than 90% of the FDA approved drugs and 75% of experimental drugs.

In another study, Gottlieb et al. (2011) utilized multiple heterogeneous sources of evidence which were integrated into a protein–protein interaction network: drug targets, drug side effects, protein sequence and GO annotations, expression profiles and disease phenotype data. They defined several profile-based similarity measures for drugs and diseases: chemical structure based, protein and genetic sequence based, phenotype based, side effect based, network topology based and GO annotation based. The similarities measures were subsequently combined into association scores and used as features for a logistic regression classifier to identify novel drug indications.

Lee et al. (2012) constructed a tripartite drug–protein–disease network based on a large integrative database incorporating drug targets, disease-associated proteins, protein interaction, and pathway data. To explore drug–disease associations within the network they used an in-house algorithm called shared neighborhood scoring. This algorithm allowed them to predict drug–disease pairs based on the guilt by association principle that unlinked pairs which share significant numbers of neighbors with strong relationships between them could be confidently linked. They used this approach to suggest as repositioning candidate for lung cancer treatment the high blood pressure drug benzthiazide.

Zhao and Li (2012) also used a drug–protein–disease network and developed a Bayesian partition method to retrieve drug–protein–disease modules which were closely connected. The authors started from a comprehensive protein–protein interaction network assembled by integrating data from several databases. Subsequently, information from disease–genes relations from OMIM and drug–target interactions from DrugBank were mapped onto the protein–protein interaction network. Then, gene–drug paths were computed to reflect the network distance between a gene and each drug's targets. Similarly, gene–disease closeness was estimated to reflect the network distance between a gene and each disease-related genes. Based on these network distances drug-gene-disease modules were identified using a Bayesian partition method. The approach was used to infer drug–disease associations, and suggest new drug applications for anti-asthma drug pranlukast (repositioned for treating cancer metastasis) and cardiovascular stress-testing agent arbutamine (repositioned for treatment of obesity).

Based on the same strategy of searching for closely connected modules (whose members are more likely to be functionally related) within drug–protein–disease networks, Daminelli et al. (2012) implemented a method that searches the network for bi-cliques motifs. In their case bi-cliques are subnetworks in which every drug is linked to every target and disease. They initially built large bipartite networks from various public databases in which drugs, targets, and diseases are linked by drug–target associations and drug-disease associations. Subsequently, network analysis based on power graphs was employed to search for incomplete cliques in the network. Bi-cliques connected by common drugs are thus identified from the bipartite network. Consequently, resulting incomplete bi-cliques' completion is used to predict novel links from drugs to targets and diseases, respectively, thus allowing the authors to simultaneously suggest reposition for drugs and predict a drug's off-targets. The approach allowed the authors to suggest and computationally validate repositioning for nine cardiovascular drugs for treating parasitic diseases.

Other approaches on drug repositioning based on molecular profiles are those of Iorio et al. (2010) who, as presented in Sect. 2.2, built a drug–drug network in which drug nodes were linked based on similarity measures derived from ranked gene lists. Their work, developed primarily for MoA discovery, suggested that fasudil, a vasodilator used in stroke, would be effective in treatment of autophagy, which is a major process in cancer. Another work based on molecular profiles and which links MoA to drug repositioning is that of Iskar et al. (2013), also described in Sect. 2.2 above, which identified conserved drug-induced modules from transcriptional profile data and enriched the modules with information from various other databases. Module membership was then used to induce novel indications for existing drugs, predictions which were further validated experimentally. Vasodilator vinburnine, topical antifungal sulconazole, and cardiac stimulant mephentermine were all suggested as candidates as cell-cycle inhibitors in anti-cancer therapy.

In a very recent work Guney et al. (2016), propose an innovative approach which transcends the drug repositioning area, having possible applications also in drug MoA elucidation as well as drug-target identification areas. The authors introduce the concept of drug–disease proximity based on shortest paths between target and disease associated genes within the interactome. They argue that proximity to disease small neighborhoods is a good proxy for describing therapeutic effect and improve the accuracy of drug repositioning predictions. Following this approach they explain why HIV drug plerixafor is repurposed for non-Hodgkin' lymphoma and provide potential repositioning candidates for rare diseases.

A recent trend is the use of noncoding RNAs, such as miRNA, as therapeutic agents due to their regulation of cellular processes implicated in disease. As a consequence, drug repositioning strategies considering miRNAs are also attracting significant interest. Liu et al. (2014), devised an approach for identifying repositioning candidates for cystic fibrosis based on miRNA-transcription factors feed forward loops. The loops are essentially motifs in a regulatory network with connectivity patterns occurring more frequently than in control networks, and therefore could be seen as response subnetworks. Using GEO expression data, gene-miRNA relationship data, protein interaction, and drug-miRNA interaction data as well as disease-related gene data from public databases, they built regulatory networks which were searched for feed forward loops implicated in cystic fibrosis. They found 48 drugs showing ability to perturb the expression of miRNAs which are part of loops implicated in cystic fibrosis, and which were suggested for repositioning. Similarly, Jiang et al. (2012) have developed a method that searches for modules in a drug-miRNA human cancer network built from CMap data, miRNA target gene databases and enriched with GO annotations. Using hypergeometric tests on the retrieved modules they suggested that 2-deoxy-D-glucose (2DOG) is a candidate for treating thyroid cancers.

2.4.2 Drug Repositioning Based on Phenotypic Profiles

Drug repositioning approaches based on phenotypic profiles typically rely on the principle that, if a drug shares similar side effect profile with a set of drugs prescribed to treat a specific disease, then the respective drug can be considered as a candidate for treating that disease (Wu et al. 2013). Since drug side effects are usually generated when drugs bind to off-targets (known or unknown), and hence perturb metabolic or signaling pathways, it is expected that the side effect profile of drugs may reveal relevant unknown information pertaining their MoA, and hence assist in repositioning.

One of the first works following this principle was that of Campillos et al. (2008) which, as already described in Sect. 2.3.2 used a side effect similarity profile incorporated into a drug–target network to infer probability of two drugs sharing the same target. Based on this, authors identified phenotypic associations between nootropic drug donepezil and antidepressant venlafaxine and suggested a new market use for donepexil in treating depression.

In another work, Yang and Agarwal (2011) used a disease–side effect network by combining drug-disease associations with drug–side effect associations from PharmGKB and SIDER databases, respectively. Subsequently, they used Naïve Bayes predictors trained on relations between side effect and disease for predicting new indications for drugs. Following this approach they predicted that drugs associated with increased immune response, such as ticlopidine and ACE inhibitors are potential candidates for treating stroke. Ye et al. (2014) used a side effect-based similarity measure to connect drugs into a drug–drug network and searched for subnetwork neighborhoods enriched with drugs having a specific therapeutic indication. They used the guilt by association approach to assign a new indication to drugs present in the same subnetwork. They suggested a number of candidate drugs for repositioning, among which the analgesic drug tramadol and Parkinson's drug tolcapone in treating depression.

One of the problems related to guilt by association approaches is that they often enforce restrictions on the search space by only considering most similar drug, discarding possible useful information embedded in the whole dataset. Bisgin et al. (2014) used the assumption that all phenotypes in the phenome (both drug indications and side effect) are interconnected with a probabilistic distribution and used a probabilistic generative model for their analysis. They used a Bayesian based model, the Latent Dirichlet Allocation (LDA) to uncover links between drugs and phenotypes, which are actually novel indications. Links are encoded into conditional probabilities. Although their method does not explicitly make use of biological networks, the LDA model they chose can be represented as a tripartite network constructing paths from drugs to phenotypes via connections across latent variables. They suggested new treatment options for all 908 drugs in their study, among which some were confirmed by literature validation, e.g., influenza A drug amantadine's use for treating epilepsy.

Finally, we must note the development in the recent years of several web servers and open-source packages for the specific goal of drug repositioning, which integrate resources covering both molecular profile and phenotype-based approaches. With some variations, they all rely on the integration of heterogeneous data sources to build the interactome network, and incorporate some of the previously published similarity measures. Among the most popular are the PROMISCUOUS (Van Eichborn et al. 2011), DRAR-CPI (Luo et al. 2011), DMAP (Huang et al. 2015) and ksRepo (Brown et al. 2016). PROMISCUOUS integrates relations between drugs, targets, and side effects and uses drug structural similarity and side effect similarity measures. It allows users to search by single drug ID queries or perform network-based exploration given a set of drugs and targets. DRAR-CPI only uses chemical structure in the chemical-protein interactome to predict network based drug–drug associations and produce lists of drugs which share similar interaction profiles and side effect information with the query drug. DMAP combines both chemical-protein interactome, protein–protein interactions, transcriptional profiles, and phenotype data (disease indications) to build a directional weighted interactome network. They use already published gene similarity (Iorio et al. 2010) and drug similarity measures to derive a guilt by association model based on the Kolmogorov–Smirnov enrichment (Lamb et al. 2006) to predict novel indications for drugs. ksRepo is a recent open-source software package implemented in R which proposes a generalized methodology enabling integration of transcriptional profiles from various platforms (including RNA-seq). Their method is based on disease transcriptional profiles and gene–drug interactions (available from any user desired source). They implement a variant of the Kolmogorov–Smirnov enrichment to compare single instances (disease transcriptional profile) with multiple drug–gene interaction lists and then derive scores which reflect disease–drug associations based on the transcriptional profiles.

2.5 Network-Based Side Effect Modeling and Prediction

Drug side effects are among the most important factors to be considered in drug design. Recent studies estimated side effects to be the major reason for drug discontinuation in first phase clinical trials and second most common cause of drug attrition overall (Hornberg et al. 2014). Therefore, computational approaches for in silico prediction of side effects are highly relevant, and currently under consideration by the pharmaceutical industry in their effort to complement the high throughput in vitro screening of newly developed drugs (Bowes et al. 2012).

Side effects are the result of promiscuous binding behavior of the majority of drugs, which in addition to their primary targets can interact with different affinities with many off-targets (Paolini et al. 2006). This way they potentially perturb many signaling and metabolic pathways eliciting both therapeutic effects and unwanted physiological responses. These signaling and metabolic pathways are often partially overlapping, thus producing synergistic or canceling consequences. Currently, there are several important observations and hypotheses which guide research in this area: different drugs can share similar side effect profiles as a result of sharing similar toxicological pathways or networks, which is an extension of the observation that the result of drug on-target and off-target binding behavior is a perturbation that is relayed downstream to partially overlapping (cross-talking) pathways (Bai and Abernethy 2013). This is related also to the principle which states that if a drug shares similar side effect profile with a set of drugs prescribed to treat a specific disease, then the respective drug can be considered as a candidate for treating that disease (Wu et al. 2013). The recent observation that network neighborhood of drug targets is a major determinant of side effect similarity profiles of drugs comes as a corollary to the previously enounced principles (Browers et al. 2011). Consequently, the development of in silico methods for side effect prediction is significantly benefiting from the increased interest in the area of drug–target prediction.

The computational approaches based on network analysis aiming at predicting drug side effects and modeling their generation mechanisms can be broadly categorized as being chemical-based and pathway-based. Both types of approaches heavily use the two important concepts in network modeling: that of network neighborhood (which define areas of the network with inter-related and coherent functional properties), and that of similarity (which is defined on various chemical, genomic or ontology features to reflect proximity between network nodes or neighborhoods).

2.5.1 Approaches Based on Chemical Structure

Chemical-based approaches generally attempt to relate chemical structure of drugs to their side effects, based on the basic observation that similar ligands interact with similar proteins. Thus, based on the backbone consisting of drug chemical structure, protein structure and information on drug–target interactions and incomplete drug– side effect association, models can be built to predict novel drug–side effect associations. Some examples include the work of Schreiber et al. (2009) which developed a method integrating various sources on chemical substructures and information on side effects to find large-scale structure–side effect associations. In their network they linked side effects based on correlations between drug chemical features. Their aim was not a mechanistic understanding of side effect causes but rather drawing a global picture of how different types of side effects may be linked, with the goal of defining possible filters for screening drug compound candidates. Similarly, Pauwels et al. (2011) used sparse canonical correlation analysis (SCCA) to predict side effects and associate them with correlated ensembles formed by chemical substructures. Yamanishi et al. (2010) proposed a unified framework, based on the integration of chemical, genomic, and pharmacological data (and the related similarity measures) with the topology of drug–target interaction networks. Within the framework of supervised bipartite network inference, using a regression approach, they were able to predict the side effect profiles of candidate drug compounds, as well as interpret drug–target interactions. In a subsequent study, they suggested several extensions to the kernel regression model for multiple responses in order to optimally integrate the heterogeneous data sources (Yamanishi et al. 2012). Based on this approach they were able to predict rare side effects for molecules in DrugBank with no available information in SIDER, such as ovarian cyst, breast tenderness, and melisma for synthetic progestational hormone drug lovonorgestrel, which were further validated based on literature.

Mizutani et al. (2012) used the co-occurrence of drugs in protein-binding profiles and side effect profiles to extract correlated sets of drug targets and side effects, using SCCA. They used a drug–target interaction network and enrichment analysis, using KEGG and GO data, to show that the retrieved correlated sets were significantly enriched in the same biological pathways, despite having different molecular functions. A biologically relevant interpretation of their results suggests that extracted side effects can be seen as possible phenotypic outcomes of drugs targeting proteins that appear in the same correlated set (i.e., having similar structures), thus reinforcing the principle mentioned in the previous paragraphs which states that target neighborhood is a predetermining factor for side effects similarity. Their side effect predictions include tremor, constipation, and dry mouth for antihistaminic drug cinnarizine, all of which were confirmed by literature of FDA reports.

Atias and Sharan (2011) combined the SCCA with a diffusion model based on side effect similarity networks. Their approach uses SCCA to project correlated structure-side effect data into a lower dimensional space. This projection is then used to predict side effects. Subsequently, using a query drug and a diffusion model on side effect similarity networks they obtain ranked list of side effects. Their validation scheme was based on a large-scale blind test based on 448 drugs from the Hazardous Substances Data Bank. The approach was able to predict correct side effects in the top 5 ranked predictions for >56% of the drugs in the database.

Lounkine et al. (2012) first used a chemical structure similarity metric, named the similarity ensemble approach (SEA), to predict targets among a set of proteins and subsequently develop a guilt by association metric that links the new targets to the side effects of the related drugs, virtually creating a drug–target–side effect network. For predicting drug target–side effect association they used an enrichment score based on co-occurrence of pairs that were more common than expected by chance, coupled with a statistically significant threshold. Based on this approach, authors predicted epigastralgia as side effect associated with chlorotrianisene, a synthetic non-steroidal estrogen. Interestingly, the off-target protein for this drug, predicted by authors, COX1, bears no sequence or structural similarity with the drug's primary target (the estrogen nuclear hormone receptor) but cross-activity between the targets is suggested by ligand similarity.

In a recent study, Wang et al. (2016) depart from the target-based approach that currently dominates the drug side effect prediction field. Their approach aims at avoiding the bias induced in the analysis by the incomplete knowledge on drug targets by combining chemical structure information with transcriptional profiles from LINCS database. They use feature sets created from signature transcriptional profiles for each drug instance, cell morphological profiles, drug chemical structure, and enrichment analysis to train a machine learning classifier based on extra trees. The most predictive classifiers are then used to shed light on the mechanisms of side effects.

Interesting insights into the factors contributing to drug side effect resulted from the approach presented in Wang et al. (2013), where authors use a structurally resolved interaction network to systematically examine relationships between drug associated side effects and drug targets. They use a generalized linear regression model and show that it is the number of essential targets (proteins which are critical for cellular survival), and not the total number of targets, that determines the side effects of drugs. Additionally, they highlight several key network topology characteristics of drug targets that are highly correlated with increased side effects profiles. They noted that high node degree (number of interactions for a target) and betweenness (the number of shortest paths between other proteins in the network passing through the target protein), as well as highly shared interaction profiles are more likely to result in an increase in the number of side effects.

2.5.2 Approaches Based on Pathways/Sub-pathways

Pathway-based approaches relate drug side effects to perturbed biological pathways or sub-pathways which contain drug target proteins. Consequently, they train models on the molecular interaction networks built from various data sources (such as drug–target interactions, gene/protein–disease—drug–side effect connections, or drug–drug interactions) in order to predict side effects for unknown drug–side effect associations based on underlying network motifs. The models thus derived are able to provide mechanistic insights into the side effect generation process.

Lee et al. (2011) used an enrichment score to define drug-biological process associations based on CMap transcriptional profiles and GO ontologies and subsequently built multilevel biological process–drug–side effect network to discover relationships between biological processes and side effects, using drug information as a bridge. For this purpose they employed a co-occurrence-based scoring accounting for how many drugs shared the same side effect in a specific biological process. Bauer-Mehren et al. (2012) use a two-step framework for biological annotation of side effects with relevant pathways. They search for drug–target and target–side effect associations and then compare these associations to derive drug–side effect links. In a subsequent step they substantiate the found associations using pathway information from Reactome database.

Li et al. (2012) used a bipartite drug-metabolic sub-pathway network build after identifying sets of drug-induced differentially expressed genes from CMap and pathway enrichment analysis. By analyzing drug–sub-pathway associations they uncovered that drugs share similar indications and side effect if they are associated to same sub-pathways. Additionally, an increase in the number of sub-pathways shared by drugs correlates with increased numbers of common side effects. Overall, their study confirms the idea that important therapeutic and side effect related mechanisms are relayed through sub-pathways, which are smaller regions of pathways, and may be overlooked by whole pathway-based methods. In a related study highlighting the importance of subnetwork-based approaches, Zhao et al. (2013), proposed an approach for identifying drug combinations to mitigate side effects. To this goal they used a human interactome network built from protein–protein interaction databases and then searched for subnetworks enriched with sets of related GO biological processes annotations. Interactions between drug pairs based on their targets were searched using a random walk method and correlated with information on their side effects. As mentioned in Sect. 2.3.3, following this approach they were able to predict the mitigating effect of exenatide on rosiglitazone's myocardial infarction side effect and explain that this could occur through a clotting regulation mechanism.

Another subnetwork-based approach was followed by Lorberbaum et al. (2015) which also used an initial interactome network created from protein–protein interactions. Their initial network was pruned based on data from several sources and biological levels, such as to highlight subnetwork modules with mechanistic connections to phenotypes. Their subnetworks were enriched in putative side effect mechanistic pathways and, subsequently, drugs were assigned to subnetworks where their targets were present. Then, subnetworks were used as features in a random forest-based classifier trained to predict whether a given drug will cause side effects.

A number of other approaches combine pathway-based analysis with information related to chemical structure of drugs and their target proteins for a holistic view on mechanisms generating side effects. Examples include the works in (Wallach et al. 2010; Fan et al. 2012), which use pathway information and in silico virtual docking to identify off-targets of drugs and link them to biological pathways. In (Liu et al. 2012), authors integrate information on drug chemical structure with pathway information and phenotypic characteristics of drugs including indications and side effect. They used a machine learning-based approach to build and evaluate the side effect prediction model. Similarly, Kuang et al. (2014) used a number of structural features of drugs integrated with network topology features of the drug–side effect association networks (constructed using correlation based methods) to build classifiers able to predict side effects.

Recently, Cao et al. (2015) integrated multiple data sources such as chemical structure, sequence, transcriptional profiles, ontology and pathways and defined multiple similarity measures based on these data types. Additionally, network topology-based similarity measures were defined, including nearest neighbor and path-based measures, using a drug–side effect network. Classification features were constructed from these similarity measures based on collaborative filtering, and a multiple evidence fusion algorithm was used for creating a multiscale predictor for side effects.

As in the case of the other application areas of systems pharmacology, a number of web servers were created for enabling the prediction of drug's side effects. The most popular among these are: IntSide (Juan-Blanco et al. 2015), which is a hybrid approach incorporating both structural and pathway information to provide mechanistic insights into drug–side effect associations. Dr. Prodis is a structure-based tool which implements several structure-pocket and structure–structure comparison procedures. Besides predicting drug side effects, it produces also drug–target interaction predictions, as well as associations between drugs and diseases (Zhou et al. 2015).

2.6 Current Challenges and Future Considerations

Despite the great promise, systems pharmacology approaches face a number of challenges while scaling from pre-clinical setting into clinical applications. A major hurdle is the bias caused by incomplete knowledge. For example, network-based models tend to bias to the targets with more known associated drugs, and even if current studies, such as the one in Wang et al. (2016) attempt to balance their models by incorporating sources of evidence from different biological levels, such as chemical structure, lack of adequate high resolution structural data for targets may induce further problems. However, recent progress in both experimental and computational methods in the area of structural genomics holds the promise to significantly improve the structural coverage. Another limitation of almost all network-based approaches, especially those relying on searching paths across the network is that they cannot provide predictions (e.g., for drug–target pairs) when missing information hampers the identification of reachable paths in the network. Network-based methods need to adequately address these aspects in the future.

Another important issue is that of the amounts of data at multiple scales needed to build accurate predictive models in the context complex disease heterogeneity. And whether the incorporation of such specialized data will still produce models with decent generalization performance, given for example an individual with unobserved new mutation. Current approaches treat insufficiently the problem of inter-subject genetic variability, which is a crucial step toward the goal of precision medicine. Among the other challenges worth mentioning are the lack of structured gold standard, especially in the applications related to drug repositioning and side effect prediction. Ideally, in silico experimental results should be integrated into the drug design validation pipeline and tested in binding experiments, cellular assays or animal models for not only providing filters for initial candidate lists, but also retrieving false positives that could be further used to refine the algorithms. While in the case of drug-target prediction and MoA characterization, the results provided by predictive models can be easily tested experimentally, for drug repositioning and side effect prediction it is often the case that genomic responses in animal models vary significantly when compared to human models. Therefore, additional care must be taken for thorough training and testing of the predictive models. From this perspective, the availability of extensive secondary use data from patients electronic health records, presents researchers valuable resources for performing 'retrospec-tive' experiments on human subjects in clinical settings (Lorberbaum et al. 2016).

Another aspect is that, despite the increased predictive power generated by the incorporation of multiscale heterogeneous data into the network and statistical models, there still is the question of how relevant it is to discover new knowledge from static statistical models, under conditions that are constantly changing. Under drug treatment, a disease state is not static, but evolves through successive states while responding to the drug-induced perturbations. When sufficient data are col-lected to successfully build a model describing one disease state, the disease may already be in a different state from the one used to build the model. In such a dynamic situation, a data-driven model is essentially retrospective and not prospective (Xie et al. 2014). Presently, very few methods that offer dynamic resolution are used in systems pharmacology. One such approach is that of (Bansal et al. 2006) where an algorithm called Time Series Network Analysis (TSNI) was proposed to infer targets of antibiotic norfloxacin based on time series transcrip-tional profiles experimental data. As more time course experimental data is bound to be produced, dynamic methods such as CHRONOS (Vrahatis et al. 2016a), described in Chap. 3 could become prevalent. CHRONOS can be easily tailored to provide a framework for studying sub-pathways activated by drugs, or other therapeutic molecules, at specific drug treatment stages. Additionally, such an approach could be adapted to identify sub-pathways perturbed during disease progression and optimal time points for drug treatment could be inferred.

The evolution of network and pathway-based approaches used in systems pharmacology has followed a journey starting with methods based on overrepresentation, or enrichment analysis, which use statistical tests to find sets of genes in a particular pathway among the (usually differentially expressed) genes under study (Khatri et al. 2012). A second generation of approaches uses gene-level statistics (e.g., identifying individual differentially expressed genes) and then aggregating the gene-level statistics of all genes in a pathway into a pathway-level statistic or score, such as the Kolmogorov–Smirnov statistic (Lamb et al. 2006; Iorio et al. 2010, 2015). The third generation of methods are topology based, thus allowing the incorporation of information from various sources, beyond simple lists of genes (Pan et al. 2014; Wallach et al. 2010; Fan et al. 2012). We believe the trend in recent publications, which consider small regions of pathways (sub-pathways) with specific topological features and which are activated by perturbations cause by disease or drug treatment, is likely to follow and further expand. The development of robust sub-pathway-based approaches able to provide useful insights into time- and condition-specific activated sub-pathways (CHRONOS), and assist in identifying disease perturbed sub-pathways [DEsubs (Vrahatis et al. 2016b), described in Chap. 4] is therefore of utmost importance for future studies in systems pharmacology.

References

Ambesi-Impiombato A, Diego d Bernardo A (2005) Computational biology and drug discovery: from single-target to network drugs. Curr Bioinf 1:3–13

Anighoro A, Bajorath J, Rastelli G (2014) Polypharmacology: challenges and opportunities in drug discovery: miniperspective. J Med Chem 57:7874–7887

Antman E, Weiss S, Loscalzo J (2012) Systems pharmacology, pharmacogenetics, and clinical trial design in network medicine. Wiley Interdisc Rev Syst Biol Med 4:367–383

Arrell DK, Terzic A (2010) Network systems biology for drug discovery. Clin Pharmacol Ther 88:120–125

Atias N, Sharan R (2011) An algorithmic framework for predicting side effects of drugs. J Comput Biol 18:207–218

Bai JP, Abernethy DR (2013) Systems pharmacology to predict drug toxicity: integration across levels of biological organization. Annu Rev Pharmacol Toxicol 53:451–473

Bansal M, Della Gatta G, Di Bernardo D (2006) Inference of gene regulatory networks and compound mode of action from time course gene expression profiles. Bioinformatics 22:815–822

Barabási AL, Gulbahce N, Loscalzo J (2011) Network medicine: a network-based approach to human disease. Nat Rev Genet 12:56–68

Bashan A, Bartsch RP, Kantelhardt JW, Havlin S, Ivanov PC (2012) Network physiology reveals relations between network topology and physiological function. Nat Comm 3:702

Bauer-Mehren A, Van Mullingen EM, Avillach P et al (2012) Automatic filtering and substantiation of drug safety signals. PLoS Comput Biol 8:e1002457

Berger SI, Iyengar R (2009) Network analyses in systems pharmacology. Bioinformatics 25:2466–2472

Bezerianos A, Maraziotis IA (2008) Computational models reconstruct gene regulatory networks. Mol BioSyst 4:993–1000

Bisgin H, Liu Z, Fang H, Kelly R, Xu X, Tong W (2014) A phenome-guided drug repositioning through a latent variable model BMC. Bioinformatics 15:1

Boran AD, Iyengar R (2010) Systems approaches to polypharmacology and drug discovery. Curr Opin Drug Discov Devel 13:297

Bowes J, Brown AJ, Hamon J, Jarolimek W, Sridhar A, Waldron G, Whitebread S (2012) Reducing safety-related drug attrition: the use of in vitro pharmacological profiling. Nat Rev Drug Discov 11:909–922

Brouwers L, Iskar M, Zeller G, Van Noort V, Bork P (2011) Network neighbors of drug targets contribute to drug side-effect similarity. PLoS ONE 6:e22187

Brown AS, Kong SW, Kohane IS, Patel CJ (2016) ksRepo: a generalized platform for computational drug repositioning. BMC Bioinf 17:1

Campillos M, Kuhn M, Gavin AC, Jensen LJ, Bork P (2008) Drug target identification using side-effect similarity. Science 321:263–266

Cao DS, Xiao N, Li YJ, Zeng WB et al (2015) Integrating multiple evidence sources to predict adverse drug reactions based on a systems pharmacology model. CPT: Pharmacometr Syst Pharmacol 4:498–506

Chen X, Xu J, Huang B, Li J et al (2011) A sub-pathway-based approach for identifying drug response principal network. Bioinformatics 27:649–654

Chen X, Liu MX, Yan GY (2012) Drug–target interaction prediction by random walk on the heterogeneous network. Mol BioSyst 8:1970–1978

Chen X, Yan CC, Zhang X, Zhang X, Dai F, Yin J, Zhang Y (2015) Drug–target interaction prediction: databases, web servers and computational models. Briefings Bioinf bbv066

Cheng F, Liu C, Jiang J, Lu W et al (2012a) Prediction of drug-target interactions and drug repositioning via network-based inference. PLoS Comput Biol 8:e1002503

Cheng F, Zhou Y, Li W, Liu G, Tang Y (2012b) Prediction of chemical-protein interactions network with weighted network-based inference method. PLoS ONE 7:e41064

Csermely P, Korcsmaros T, Kiss HJ, London G, Nussinov R (2013) Structure and dynamics of molecular networks: a novel paradigm of drug discovery: a comprehensive review. Pharmacol Ther 138:333–408

Daminelli S, Haupt VJ, Reimann M, Schroeder M (2012) Drug repositioning through incomplete bi-cliques in an integrated drug–target–disease network. Integr Biol 4:778–788

DiMasi JA, Grabowski HG, Hansen RW (2016) Innovation in the pharmaceutical industry: new estimates of R&D costs. J Health Econ 47:20–33

Esserink M (2016) Science Magazine. doi:10.1126/science.aaf4017, http://www.sciencemag.org/news/2016/02/french-company-bungled-clinical-trial-led-death-and-illness-report-says. Retrieved Oct 2016

Fan S, Geng Q, Pan Z, Li X, Tie L, Pan Y, Li X (2012) Clarifying off-target effects for torcetrapib using network pharmacology and reverse docking approach. BMC Syst Biol 6:152

FDA (2016) Table of pharmacogenomic biomarkers in drug labeling. http://www.fda.gov/Drugs/ScienceResearch/ResearchAreas/Pharmacogenetics/ucm083378.htm. Retrieved 5 Oct 2016

Ganter B, Tugendreich S, Pearson CI et al (2005) Development of a large-scale chemogenomics database to improve drug candidate selection and to understand mechanisms of chemical toxicity and action. J Biotechnol 119:219–244

Gardner TS, Di Bernardo D, Lorenz D, Collins JJ (2003) Inferring genetic networks and identifying compound mode of action via expression profiling. Science 301:102–105

Gene Ontology Consortium (2013) Gene ontology annotations and resources. Nucleic Acids Res 41(D1):D530–D535

Glaab E, Baudot A, Krasnogor N, Valencia A (2010) Extending pathways and processes using molecular interaction networks to analyze cancer genome data. BMC Bioinf 11:1

Gottlieb A, Stein GY, Ruppin E, Sharan R (2011) PREDICT: a method for inferring novel drug indications with application to personalized medicine. Mol Syst Biol 7:496

Graham DJ, Campen D, Hui R et al (2005) Risk of acute myocardial infarction and sudden cardiac death in patients treated with cyclo-oxygenase 2 selective and non-selective non-steroidal anti-inflammatory drugs: nested case-control study. Lancet 365:475–481

Guney E, Menche J, Vidal M, Barábasi AL (2016). Network-based in silico drug efficacy screening. Nat Commun 7

Hopkins AL (2008) Network pharmacology: the next paradigm in drug discovery. Nat Chem Biol 4:682–690

Hornberg JJ, Laursen M, Brenden N, Persson M, Thougaard AV, Toft DB, Mow T (2014) Exploratory toxicology as an integrated part of drug discovery. Part I: why and how. Drug Discov Today 19:1131–1136

Huang H, Nguyen T, Ibrahim S, Shantharam S, Yue Z, Chen JY (2015) DMAP: a connectivity map database to enable identification of novel drug repositioning candidates. BMC Bioinf 16 (Suppl 13):S4

Hutchinson L, Kirk R (2011) High drug attrition rates—where are we going wrong? Nat Rev Clin Oncol 8:189–190

Hwang WC, Zhang A, Ramanathan M (2008) Identification of information flow-modulating drug targets: a novel bridging paradigm for drug discovery. Clin Pharmacol Ther 84:563–572

Hwang S, Kim CY, Ji SG et al (2016) Network-assisted investigation of virulence and antibiotic-resistance systems in Pseudomonas aeruginosa. Sci Rep 6

Ideker T, Galitski T, Hood L (2001) A new approach to decoding life: systems biology. Annu Rev Genomics Human Genet 2:343–372

Ideker T, Ozier O, Schwikowski B, Siegel AF (2002) Discovering regulatory and signalling circuits in molecular interaction networks. Bioinformatics 18(suppl 1):S233–S240

Iorio F, Bosotti R, Scacheri E et al (2010) Discovery of drug mode of action and drug repositioning from transcriptional responses. Proc Nat Acad Sci 107:14621–14626

Iorio F, Shrestha RL, Levin N, Boilot V, Garnett MJ, Saez-Rodriguez J, Draviam VM (2015) A semi-supervised approach for refining transcriptional signatures of drug response and repositioning predictions. PLoS ONE 10:e0139446

Isik Z, Baldow C, Cannistraci CV, Schroeder M (2015) Drug target prioritization by perturbed gene expression and network information. Sci Rep 5

Iskar M, Zeller G, Blattmann P et al (2013) Characterization of drug-induced transcriptional modules: towards drug repositioning and functional understanding. Mol Syst Biol 9:662

Jeong H, Tombor B, Albert R, Oltvai ZN, Barabási AL (2000) The large-scale organization of metabolic networks. Nature 407:651–654

Jiang W, Chen X, Liao M et al (2012) Identification of links between small molecules and miRNAs in human cancers based on transcriptional responses. Sci Rep 2:282

Jin G, Fu C, Zhao H, Cui K, Chang J, Wong ST (2012) A novel method of transcriptional response analysis to facilitate drug repositioning for cancer therapy. Cancer Res 72:33–44

Juan-Blanco T, Duran-Frigola M, Aloy P (2015) IntSide: a web server for the chemical and biological examination of drug side effects. Bioinformatics 31:612–613

Kanehisa M, Sato Y, Kawashima M, Furumichi M, Tanabe M (2015) KEGG as a reference resource for gene and protein annotation. Nucleic Acids Res gkv1070

Khatri P, Sirota M, Butte AJ (2012) Ten years of pathway analysis: current approaches and outstanding challenges. PLoS Comput Biol 8:e1002375

Kibble M, Khan SA, Saarinen N, Iorio F, Saez-Rodriguez J, Mäkelä S, Aittokallio T (2016) Transcriptional response networks for elucidating mechanisms of action of multitargeted agents. Drug Discov Today 21:1063–1075

Kuang Q, Wang M, Li R, Dong Y, Li Y, Li M (2014) A systematic investigation of computation models for predicting adverse drug reactions (ADRs). PLoS ONE 9:e105889

Kuhn M, Letunic I, Jensen LJ, Bork P (2015) The SIDER database of drugs and side effects. Nucleic Acids Res gkv1075

Lamb J, Crawford ED, Peck D et al (2006) The connectivity map: using gene-expression signatures to connect small molecules, genes, and disease. Science 313:1929–1935

Lázár V, Nagy I, Spohn R et al (2014) Genome-wide analysis captures the determinants of the antibiotic cross-resistance interaction network. Nat Commun 5

Lee S, Lee KH, Song M, Lee D (2011) Building the process-drug–side effect network to discover the relationship between biological processes and side effects. BMC Bioinf 12:1

Lee HS, Bae T, Lee JH et al (2012) Rational drug repositioning guided by an integrated pharmacological network of protein, disease and drug. BMC Syst Biol 6:80

Li C, Shang D, Wang Y, Li J et al (2012) Characterizing the network of drugs and their affected metabolic subpathways. PLoS ONE 7:e47326

Li ZC, Huang MH, Zhong WQ, Liu ZQ, Xie Y, Dai Z, Zou XY (2015) Identification of drug-target interaction from interactome network with "guilt-by-association" principle and topology features. Bioinf btv695

Li J, Zheng S, Chen B, Butte AJ, Swamidass SJ, Lu Z (2016) A survey of current trends in computational drug repositioning. Briefings Bioinf 17:2–12

Liu M, Wu Y, Chen Y et al (2012) Large-scale prediction of adverse drug reactions using chemical, biological, and phenotypic properties of drugs. J Am Med Inf Assoc 19:e28–e35

Liu Z, Borlak J, Tong W (2014) Deciphering miRNA transcription factor feed-forward loops to identify drug repurposing candidates for cystic fibrosis. Genome Med 6:1

Liu X, Gao Y, Peng J et al (2015) TarPred: a web application for predicting therapeutic and side effect targets of chemical compounds. Bioinformatics btv099

Lorberbaum T, Nasir M, Keiser MJ, Vilar S, Hripcsak G, Tatonetti NP (2015) Systems pharmacology augments drug safety surveillance. Clin Pharmacol Ther 97:151–158

Lorberbaum T, Sampson KJ, Woosley RL, Kass RS, Tatonetti NP (2016) An integrative data science pipeline to identify novel drug interactions that prolong the QT interval. Drug Saf 39:433–441

Lounkine E, Keiser MJ, Whitebread S et al (2012) Large-scale prediction and testing of drug activity on side-effect targets. Nature 486:361–367

Luo H, Chen J, Shi L et al (2011) DRAR-CPI: a server for identifying drug repositioning potential and adverse drug reactions via the chemical–protein interactome. Nucleic Acids Res gkr299

Ma'ayan A, Jenkins SL, Goldfarb J, Iyengar R (2007) Network analysis of FDA approved drugs and their targets. Mount Sinai J Med J Transl Personalized Med 74:27–32

Maraziotis I, Dragomir A, Bezerianos A (2006) Gene networks inference from expression data using a recurrent neuro-fuzzy approach. In IEEE engineering in medicine and biology 27th annual conference 2006, 17 Jan IEEE, pp 4834–4837

Maraziotis IA, Dragomir A, Bezerianos A (2007) Gene networks reconstruction and time-series prediction from microarray data using recurrent neural fuzzy networks. IET Syst Biol 1:41–50

Medina-Franco JL, Giulianotti MA, Welmaker GS, Houghten RA (2013) Shifting from the single to the multitarget paradigm in drug discovery. Drug Discov Today 18:495–501

Mitsopoulos C, Schierz AC, Workman P, Al-Lazikani B (2015) Distinctive behaviors of druggable proteins in cellular networks. PLoS Comput Biol 11:e1004597

Mizutani S, Pauwels E, Stoven V, Goto S, Yamanishi Y (2012) Relating drug–protein interaction network with drug side effects. Bioinformatics 28:i522–i528

Nacher JC, Schwartz JM (2008) A global view of drug-therapy interactions. BMC Pharmacol 8:5

Napolitano F, Sirci F, Carella D, Di Bernardo D (2016) Drug-set enrichment analysis: a novel tool to investigate drug mode of action. Bioinformatics 32:235–241

Nishimura D (2001) BioCarta. Biotech Soft Internet Rep Comput Soft J Sci 2:117–120

Nissen SE, Wolski K (2007) Effect of rosiglitazone on the risk of myocardial infarction and death from cardiovascular causes. New Engl J Med 356:2457–2471

Pan Y, Cheng T, Wang Y, Bryant SH (2014) Pathway analysis for drug repositioning based on public database mining. J Chem Inf Model 54:407–418

Paolini GV, Shapland RH, van Hoorn WP, Mason JS, Hopkins AL (2006) Global mapping of pharmacological space. Nat Biotechnol 24:805–815

Pauwels E, Stoven V, Yamanishi Y (2011) Predicting drug side-effect profiles: a chemical fragment-based approach. BMC Bioinf 12:1

Pritchard JR, Bruno PM, Hemann MT, Lauffenburger DA (2013) Predicting cancer drug mechanisms of action using molecular network signatures. Mol BioSyst 9:1604–1619

Reddy AS, Zhang S (2013) Polypharmacology: drug discovery for the future. Expert Rev Clin Pharmacol 6:41–47

Rual JF, Venkatesan K, Hao T et al (2005) Towards a proteome-scale map of the human protein–protein interaction network. Nature 437:1173–1178

Scheiber J, Jenkins JL, Sukuru SCK et al (2009) Mapping adverse drug reactions in chemical space. J Med Chem 52:3103–3107

Schotland P, Bojunga N, Zien A, Trame MN, Lesko LJ (2016) Improving drug safety with a systems pharmacology approach. Eur J Pharm Sci 94:84–92

Subramanian A, Tamayo P, Mootha VK et al (2005) Gene set enrichment analysis: a knowledge-based approach for interpreting genome-wide expression profiles. Proc Nat Acad Sci 102:15545–15550

Szklarczyk D, Franceschini A, Wyder S et al (2014) STRING v10: protein–protein interaction networks, integrated over the tree of life. Nucleic Acids Res gku1003

Szklarczyk D, Santos A, von Mering C, Jensen LJ, Bork P, Kuhn M (2015) STITCH 5: augmenting protein–chemical interaction networks with tissue and affinity data. Nucleic Acids Res gkv1277

Trame MN, Biliouris, K, Lesko LJ, Mettetal JT (2016) Systems pharmacology to predict drug safety in drug development. Eur J Pharm Sci 94:93–95

Turner RM, Park BK, Pirmohamed M (2015) Parsing interindividual drug variability: an emerging role for systems pharmacology. Wiley Interdisc Rev Syst Biol Med 7:221–241

Vogt I, Prinz J, Campillos M (2014) Molecularly and clinically related drugs and diseases are enriched in phenotypically similar drug-disease pairs. Genome Med 6:1

Von Eichborn J, Murgueitio MS, Dunkel M, Koerner S, Bourne PE, Preissner R (2011) PROMISCUOUS: a database for network-based drug-repositioning. Nucleic Acids Res 39 (suppl 1):D1060–D1066

Vrahatis, AG, Balomenos P, Tsakalidis AK, Bezerianos A (2016a) DEsubs: an R package for flexible identification of differentially expressed subpathways using RNA-seq experiments. Bioinformatics btw544

Vrahatis AG, Dimitrakopoulou K, Balomenos P, Tsakalidis AK, Bezerianos A (2016b) CHRONOS: a time-varying method for microRNA-mediated subpathway enrichment analysis. Bioinformatics 32:884–892

Wallach I, Jaitly N, Lilien R (2010) A structure-based approach for mapping adverse drug reactions to the perturbation of underlying biological pathways. PLoS ONE 5:e12063

Wang X, Thijssen B, Yu H (2013) Target essentiality and centrality characterize drug side effects. PLoS Comput Biol 9:e1003119

Wang Z, Clark NR, Ma'ayan A (2016) Drug induced adverse events prediction with the LINCS L1000 data. Bioinformatics btw168

Waring MJ, Arrowsmith J, Leach AR et al (2015) An analysis of the attrition of drug candidates from four major pharmaceutical companies. Nat Rev Drug Discov 14:475–486

Woo JH, Shimoni Y, Yang WS et al (2015) Elucidating compound mechanism of action by network perturbation analysis. Cell 162:441–451

Wu Z, Wang Y, Chen L (2013) Network-based drug repositioning. Mol BioSyst 9:1268–1281

Xie L, Li J, Xie L, Bourne PE (2009) Drug discovery using chemical systems biology: identification of the protein-ligand binding network to explain the side effects of CETP inhibitors. PLoS Comput Biol 5:e1000387

Xie L, Xie L, Kinnings SL, Bourne PE (2012) Novel computational approaches to polypharmacology as a means to define responses to individual drugs. Annu Rev Pharmacol Toxicol
52:361–379

Xie L, Ge X, Tan H et al (2014) Towards structural systems pharmacology to study complex
diseases and personalized medicine. PLoS Comput Biol 10:e1003554

Xing H, Gardner TS (2006) The mode-of-action by network identification (MNI) algorithm: a
network biology approach for molecular target identification. Nat Protoc 1:2551–2554

Yamanishi Y, Araki M, Gutteridge A, Honda W, Kanehisa M (2008) Prediction of drug–target
interaction networks from the integration of chemical and genomic spaces. Bioinformatics 24:
i232–i240

Yamanishi Y, Kotera M, Kanehisa M, Goto S (2010) Drug-target interaction prediction from
chemical, genomic and pharmacological data in an integrated framework. Bioinformatics 26:
i246–i254

Yamanishi Y, Pauwels E, Kotera M (2012) Drug side-effect prediction based on the integration of
chemical and biological spaces. J Chem Inf Model 52:3284–3292

Yamanishi Y, Kotera M, Moriya Y, Sawada R, Kanehisa M, Goto S (2014) DINIES: drug–target
interaction network inference engine based on supervised analysis. Nucleic Acids Res 42:
W39–W45

Yang L, Agarwal P (2011) Systematic drug repositioning based on clinical side-effects. PLoS ONE
6:e28025

Yang K, Bai H, Ouyang Q, Lai L, Tang C (2008) Finding multiple target optimal intervention in
disease-related molecular network. Mol Syst Biol 4:228

Ye H, Liu Q, Wei J (2014) Construction of drug network based on side effects and its application
for drug repositioning. PLoS ONE 9:e87864

Yildirim MA, Goh KI, Cusick ME, Barabási AL, Vidal M (2007) Drug—target network. Nat
Biotechnol 25:1119–1126

Yuan Q, Gao J, Wu D, Zhang S, Mamitsuka H, Zhu S (2016) DrugE-Rank: improving drug–target
interaction prediction of new candidate drugs or targets by ensemble learning to rank.
Bioinformatics 32:i18–i27

Zhao S, Iyengar R (2012) Systems pharmacology: network analysis to identify multiscale
mechanisms of drug action. Annu Rev Pharmacol Toxicol 52:505

Zhao S, Li S (2012) A co-module approach for elucidating drug–disease associations and
revealing their molecular basis. Bioinformatics 28:955–961

Zhao S, Nishimura T, Chen Y et al (2013) Systems pharmacology of adverse event mitigation by
drug combinations. Sci Transl Med 5:206ra140–206ra140

Zhou H, Gao M, Skolnick J (2015) Comprehensive prediction of drug-protein interactions and side
effects for the human proteome. Sci Rep 5

Chapter 3
Time-Varying Methods for Pathway and Sub-pathway Analysis

Abstract This chapter presents in detail aspects related to pathway-based analysis of time-varying biological processes. Biological processes are inherently dynamical events involving genes and their products interacting within specific conditions. Genes are modulated by systemic perturbations (e.g., genetic modifications or drug treatments). Thus, monitoring the systemic response at multiple levels, in conjunction with the temporal evolution, is crucial to understanding and modeling the underlying biological phenomena in a comprehensive manner. The increasing need for developing biological network and pathway analysis methods capable of providing fine temporal resolution is highlighted, in the context of decreasing costs of high-throughput technologies which is expected to trigger a significant raise in time course omics experimental data availability. Several important challenges involved in this type of analysis are discussed, such as the conversion of pathway databases information into graphs (or networks) in order to allow easier interpretation of information and subsequent computational modeling, the contextualization of the transformed pathway graphs using transcriptional data, the use of search methods for the identification within graphs of paths highlighting the time dependent portions of pathways, as well as the use of various network-based statistics or interacting edge level metrics.

Keywords Pathway analysis · Linear sub-pathways · Nonlinear sub-pathways · Dynamic analysis · Time varying sub-pathways · Pathway scoring

3.1 Introduction

Pathway-based approaches were initially developed considering whole pathway features, either by means of enrichment analysis of univariate, gene level, statistics, or of global, multivariate statistics defined on the entire list of genes within the pathway. While initial results provided useful insights and aided the translation of high-throughput expression data into biologically relevant knowledge, it has been unanimously accepted that such approaches suffer from a number of shortcomings,

© The Author(s) 2017
A. Bezerianos et al., *Computational Methods for Processing and Analysis of Biological Pathways*, SpringerBriefs in Computer Science,
DOI 10.1007/978-3-319-53868-6_3

ranging from sensitivity and noise in the data to loss of information contained in the order of interactions between pathway genes, as well as disregard to the correlations and overlap between pathways (Subramanian et al. 2005; Tarca et al. 2009; Chen et al. 2011). As mentioned in the previous chapters, a recent trend in pathway analysis is that of sub-pathway-based approaches. Investigating sub-pathways is more relevant in interpreting the biological processes, since it is known that, frequently, only some regions of pathways are dysregulated by disease, or involved in drug-related perturbations, and sometimes in a transient manner. Pathways stored in biological databases are collections of specific modules (or sub-pathways), each of which are responsible for performing certain biological functions. These sub-pathways may be shared among several pathways with the goal of performing the biological role they are responsible for. Therefore, analysis considering whole pathway features is bound to overlook and miss important aspects relevant to phenotype response, whereas sub-pathway-based approaches provide the finer resolution needed to represent relevant biological processes more accurately (Chen et al. 2011).

The shifted focus toward sub-pathway-based approaches in recent years resulted in a series of methods which all share the search for targeted, context-specific, pathway portions, relevant to disease modeling, and drug targeting, namely Subpathway-GM (Li et al. 2013), TEAK (Judeh et al. 2013), DEAP (Haynes et al. 2013), *clipper* (Martini et al. 2013), and many others (Chen et al. 2011; Jacob et al. 2012; Li et al. 2012, 2015; Nam et al. 2014; Sebastian-Leon et al. 2014). The Subpathway-GM method is a representative example, in that it identifies key metabolic sub-pathways based on integrated information from genes and metabolites, taking into account their topological position within pathways and considering cascade regions within the pathways. This is achieved by searching for similarities of specific nodes within the pathway structure, termed signature nodes, and subsequently searching for shortest paths between them (Li et al. 2013). Another relevant method is the topology enrichment analysis framework (TEAK) (Judeh et al. 2013), which extracts linear and nonlinear sub-pathways, and scores them using the Bayesian Information Criterion to fit a context-specific Gaussian Bayesian network (for condition specific sub-pathways), or the Kullback–Leibler divergence (for differential sub-pathways between case and control conditions). Linear sub-pathways are extracted by exhaustively searching for root to leaf linear paths within a pathway, while nonlinear ones are found using an adaptation of the Clique Percolation Method (CPM), which allows gene nodes to participate in multiple sub-pathways and does not permit the existence of single cut nodes or cut-links within retrieved sub-pathways. DEAP (Differential Expression Analysis for Pathways—Haynes et al. 2013) exhaustively searches for paths within the pathway graphs and uses each gene node's differential expression to calculate maximum absolute running sum score along catalytic and inhibitory edges. In doing so, DEAP assumes that differential patterns of expression of paths within pathways are biologically meaningful. *Clipper* (Martini et al. 2013) follows a slightly different approach, in that it searches for paths of smaller parts (cliques) of pathways (instead of paths along gene nodes) that are relevant for signal transduction. This is achieved in a two-step approach, which first decomposes the pathway graph into a junction tree to reconstruct the most relevant signal path. In this initial step, the algorithm selects significant pathways according to statistical tests on the means and con- centration matrices of the graphs, using information from pathway topologies. In the second step, signal paths are identified within these pathways, based on their significant association with a specific phenotype.

An important aspect, which is not considered by the aforementioned sub-pathway approaches, is that biological processes are inherently dynamical events involving genes and their products interacting within specific conditions. Genes are modulated by systemic perturbations (e.g., genetic modifications or drug treatments). Thus, monitoring the systemic response at multiple levels, in con- junction with the temporal evolution, is crucial to understanding and modeling the underlying biological phenomena in a comprehensive manner. Therefore, there is an increasing need for developing biological network and pathway analysis methods capable of providing fine temporal resolution. The decreasing cost of high-throughput technologies is triggering an increase in time course omics experimental data availability, which opens new perspectives for pathway analysis. At the same time, this results in significant challenges, in terms of extending existing methods and developing novel approaches able to incorporate dynamic features of the processes under study. An additional challenge is posed by the surge in computational resources needed to accommodate this type of analyses.

Recently, a number of methods have been proposed to identify pathways and sub-pathways with significant time-dependent profiles. TimeClip (Martini et al. 2014) adapts the earlier *clipper* approach (Martini et al. 2013) to identify the most time-dependent portions of pathways. To this end, it combines principal component analysis (PCA), regression, and graph decomposition in a two-step approach: in the first step the entire pathway is searched for temporal variation by initially decomposing the data using PCA and then fitting a trend model (polynomial of degree 2) on the first principal component, whose coefficients capture existing temporal behaviors. If a certain pathway is identified as time dependent through this procedure, the algorithm decomposes it into a junction tree, having cliques as nodes, highlighting the most time-dependent portions of the pathway.

TRAP (Jo et al. 2014) proposes an approach which merges time series extensions of two classical pathway analysis approaches: over representation analysis (ORA— Goeman and Buhlmann 2007) and SPIA (Tarca et al. 2009). TRAP first finds sig- nificant pathways at each individual time point, using classical ORA (based on hypergeometric tests to identify time point specific DEGs) and SPIA (to estimate node specific perturbation factors), in order to derive a list of DEGs and of significantly perturbed pathways. In the second step, the perturbation factors found by SPIA are adapted so as to reflect the influence of upstream genes at previous time points on downstream genes at current time points. The time-dependent perturbation factors are then used as an aggregated measure to reflect the cascading effect of the perturbation throughout the entire pathway and thus provide a pathway level statistic.

TimeTP (Jo et al. 2016) narrows the focus to perturbed sub-pathways by leveraging cross-correlations between the gene expressions of nodes and uses a variation of the influence maximization algorithm, adapted from social network analysis, to evaluate the perturbation influence of certain nodes within the pathways. Their approach models the direction of perturbation propagation and the propagation delay using cross-correlation of genes' differential expression vectors. Then, paths are searched along edges which induce the shortest possible perturbation delay. Subsequently, transcription factors regulating the identified perturbed sub-pathways are considered as the seed nodes that maximize the spread of perturbation.

A recently developed method, the time-vaRying enriCHment integrOmics Sub-pathway aNalysis tOol—CHRONOS (Vrahatis et al. 2016), addresses the challenges posed by the identification of time-dependent sub-pathways from a slightly different perspective. CHRONOS models the interactions between pathway nodes based on the Markov property, meaning sub-pathway activity at time point t is depended solely on the state at time $t - 1$, and on the assumption that different portions of pathways are active at different time points. In doing so, CHRONOS defines two scoring schemes which are aimed at capturing time-dependent changes in expression level of genes. The two measures are then unified into a single score which can be used to characterize interaction, or perturbation, trends within pathways. Linear and nonlinear sub-pathways (k-cliques) are exhaustively searched within the pathway graph and ranked according to the score described above. Inter-pathway dependency is then accounted for by defining sub-pathway-based topology metrics which inform on the position of sub-pathways within the entire pathway graph. Additionally, CHRONOS searches for potential miRNA regulators of sub-pathways, by estimating the enrichment of miRNA targets within the identified sub-pathways using the cumulative hypergeometric distribution.

The majority of currently existing time dependent pathway and sub-pathway analysis methods rely on a succession of steps such as (i) conversion of pathway databases information into graphs (or networks) which allow to easier interpret information representation and subsequent computational modeling (described in Sect. 3.2), (ii) contextualization of the transformed pathway graphs using expression data, (iii) search within graphs for paths which highlight the time-dependent portions of pathways (Sect. 3.3), a step which is commonly accompanied by (iv) the use of various node or interacting edge level metrics or measures (Sect. 3.4).

3.2   Conversion of Pathway Databases to Graphs

A critical step in pathway analysis methods, and especially for approaches which consider pathway topology information, is the creation of a background graph containing a comprehensive representation of information stored in curated biological repositories, such as KEGG, PathwayCommons, ConsensusPathDB, or PANTHER. This step is used by all pathway analysis methods and it is therefore common that time-dependent methods adopt the existing conversion methodology

```
(a) <entry id="7" name="hsa:5770" type="gene"
    <entry id="36" name="hsa:3643" type="gene"
    <relation entry1="7" entry2="36" type="PPrel">
        <subtype name="inhibition" value="--|"/>
        <subtype name="dephosphorylation" value="-p"/>
    </relation>

(b) <entry id="179" name="cpd:C00078" type="compound"
    <entry id="172" name="cpd:C02700" type="compound"
    <reaction id="72" name="rn:R00678" type="irreversible"
        <substrate id="179" name="cpd:C00078"/>
        <product id="172" name="cpd:C02700"/>
    </reaction>
```

Fig. 3.1 Examples of metabolic and non-metabolic pathway entries in KEGG. **a** An entry from insulin signaling pathway (hsa04910), a non-metabolic pathway, as depicted in a KGML file, between genes with Entrez Ids '5570' and '3643', respectively. Relation type 'PPrel' indicates a protein–protein interaction between proteins and subtype 'inhibition' indicates the inhibition of node 'entry2' from 'entry1,' and at the same time the posttranslational modification taking place (dephosphorylation). **b** An entry from tryptophan metabolism pathway (hsa00380), a metabolic pathway, denoting a reaction with substrate 'C00078' and product 'C02700'

developed for static approaches. The most common repository currently used is the KEGG database, which is the most comprehensive database of metabolic and non-metabolic pathways. Pathway information in KEGG is represented using the KEGG Markup Language (KGML), which is an exchange format originally created for enabling automatic drawing of KEGG pathways. KGML provides an easily interpretable format for computational analysis and modeling of gene/protein networks and chemical networks. The KGML files for metabolic pathway maps contain two types of graph object patterns, boxes (enzymes and metabolites), and circles (chemical compounds). Enzymes are linked by 'relations' and chemical compounds are linked by 'reactions' (Fig. 3.1). Non-metabolic pathway maps in KGML contain information on how protein nodes are linked by 'relations' (Kanehisa et al. 2015). There are four main relation types, the (i) protein–protein relation (type = 'PPrel'), the (ii) gene expressions relation (type = 'GErel'), the (iii) protein-compound relation (type = 'PCrel') and, (iv) links to other maps (type = 'maplink'). From these elements many subtypes arise. The only relation type among enzymes is the enzyme–enzyme relations (type = 'ECrel'). Metabolites are mainly participating in chemical reactions with sub-element substrates and products. Detailed information on data types used by KGML is provided in http://www.genome.jp/kegg/xml/docs/.

A common first step is to convert KEGG pathways to graphs in the form of directed gene–gene networks based on information in KGML files, without compromising the structural topology of the pathway. The KEGG pathway entry elements representing genes (type = 'gene') may in some cases correspond to single or multiple gene products containing most likely gene families, or genes with similar biochemical function. Similarly, group entries (type = 'group') usually denote multiple gene products, commonly representing protein complexes. This case is modeled in pathway graphs in the following manner: the corresponding entries are expanded into separate nodes by rewiring the incoming and outgoing links of the entry (Fig. 3.2). The approach is implemented in the Bioconductor package

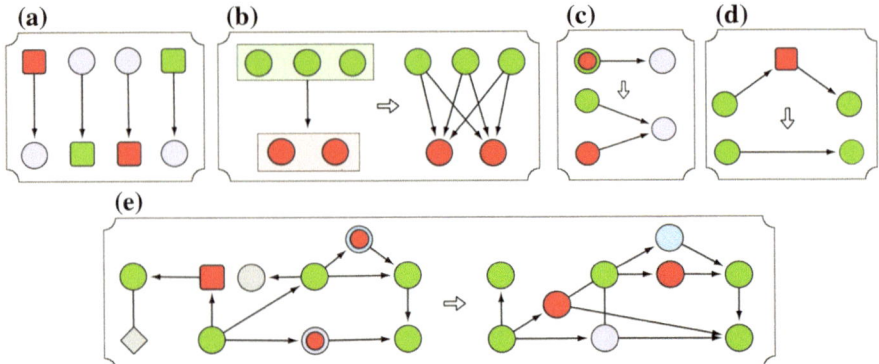

Fig. 3.2 Conversions performed during the construction of a pathway network. *Single circles* denote genes/gene products, *overlapping circles* multiple nodes, *squares* chemical compounds and *rhombuses* denote links to other pathways. **a** Edges created in a reversible reaction in metabolic pathways. If the reaction is irreversible, only the first two are created. **b** Conversion of the interaction between two groups of nodes to interactions between interconnected single nodes. **c** Conversion interaction between a multiple node and a single node to interactions between noninterconnected single nodes. **d** Removal of chemical compound, preserving the network connectivity. **e** Conversion of a pathway to a network of single nodes corresponding to genes/gene products, after performing conversions **a, b, c, d**

graphite and in the online tool Graphite Web (Sales et al. 2012, 2013), and is the approach used by methods such as CHRONOS and Subpathway-GM. A slightly different approach is followed in timeClip, where protein complexes are expanded into cliques, such that all individual proteins are connected to each other). This is done in order to assist in subsequent steps, as timeClip is a method which searches for paths formed of cliques, as subsequently described in Sect. 3.3. A different approach, which consider protein complex entries as single nodes by unifying the corresponding gene products for downstream analysis is considered by TEAK.

The chemical compounds (type = 'compound') are usually omitted when representing KEGG pathways into graphs and the pathway structure is preserved by connecting the nodes in cases a compound acts as a bridge in-between gene nodes (Fig. 3.2). This is done because compounds are not usually assessable using high-throughput technologies. However, in order to follow the biological model adequately, the compound-mediated interactions between two nodes are preserved if the bridging compound has the same cellular localization with both interacting nodes, i.e., bridges two successive reactions. (Sales et al. 2013).

Entry elements representing KEGG orthologs (type = 'ortholog') and maps (type = 'map' are usually ignored since generally each organism and pathway is analyzed separately (Fig. 3.2). It must be noted at this point that approaches based on the metabolic pathways, such as Subpathway-GM and TEAK take into consideration the type (type = 'compound') and use them for defining the metabolites as nodes in the graph.

The edges (or interactions) between nodes in the pathway graphs are defined by taking into consideration the directionality and the type of operation for each relation. In a KEGG pathway map, a relation with type = 'inhibition' between 'entry1' and 'entry2' encodes an interaction which denotes that 'entry1' inhibits 'entry2' (Fig. 3.1). However, in some cases, relation types have ambiguous interpretation, both in terms of directionality and in their operation. For example, the relation type = 'binding/association' has no specific directionality and it is commonly supposed that this relation is bidirectional. A relation of type 'activation/inhibition' has ambiguous operation. Therefore, existing approaches, such as CHRONOS, Subpathway-GM and TEAK use a categorization with three kinds of relations (activation, inhibition, unknown). Activation/expression is translated as enhancement of gene regulation, inhibition/repression as suppression of expression and unknown is used to denote interactions with unclear molecular context. Also, a fourth type of relation, indicated as 'no-interaction', can be provided as option to the user, in case there is interest to focus only on specific interaction types, by setting all the irrelevant interaction types as 'no-interaction.' The translation scheme used by CHRONOS and TEAK is based on (Wrzodek et al. 2013). An indicative example showing a general pathway conversion to gene–gene network is shown in Fig. 3.2.

In approaches which are based on metabolic pathways the main difference is that the edges are mainly constructed from chemical reactions (type = 'reaction'), which involve enzymes (gene products) or compounds (i.e substrates and products). For each interaction, the substrate, the product, and the reaction type (reversible or irreversible) are examined. More specifically, for an enzyme e and a reaction with substrate id = 's,' product id = 'p,' and type = 'irreversible,' edges (s, e) and (e, p) are created. In reversible reactions, edges (e, s) and (p, e) are created additionally (Fig. 3.2). For enzyme e and reaction with substrate id = 's,' product id = 'p,' and type = 'reversible,' edges (s, e), (e, p), (e, s), and (p, e) are created. The directionality in the two reaction types is also user defined. Reactions and relations in metabolic pathways are considered to have ambiguous operation (Büchel et al. 2013; Cicek et al. 2014) and can be defined as 'unknown' relation type.

The KGML format, the most common format due to its use by KEGG, has significant drawbacks, in that it was initially designed for visualization purposes and therefore often omits important details, resulting in ambiguities in the stored information (Wrzodek et al. 2013). Other pathway databases, such as Pathway Commons, ConsensusPathDB, WikiPathways, PANTHER, or Reactome offer support for exporting data into BioPAX (a RDF/OWL-based standard language) and/or SBML (a representation format based on XML), which are the main standardization efforts for the representation of biological pathway information. Several tools have been implemented to provide adequate translations between KEGL and SBML or BioPAX, such as the KEGGTranslator (Wrzodek et al. 2013), KEGGconverter (Moutselos et al. 2009), or Sybil (Gelius-Dietrich et al. 2013).

3.3   Time-Varying Pathway and Sub-pathway Extraction

Once pathway information located in pathway databases is converted to a format suitable for subsequent network-based analysis, the next step is to search for portions of pathways that are most time dependent. This is particularly relevant for studies which aim to investigate systemic modulations induced by perturbations, such as drug treatment, genetic modifications, or disease progression. Most commonly, prior to this step, or in some cases concomitant with the analysis, high-throughput molecular information (usually gene expression data from microarray or RNA-seq), is overlaid to the pathway topology.

Two types of approaches are followed normally: extraction of linear or nonlinear sub-pathways. The former are comprised of genes in linear cascades (paths), an approach followed by CHRONOS, TimeTP, and, to some extent, TRAP. The latter consists of extracting highly connected gene communities (cliques), an approach followed by timeClip. Some methods, such as CHRONOS, are able to identify both types of sub-pathways, or propose a hybrid approach, such as timeClip, which finds linear paths of time dependent cliques.

3.3.1   Linear Sub-pathway Extraction

If G is a pathway graph, start and end nodes are defined as follows: S_n are nodes with zero in-degree (no incoming connecting edges) and D_m nodes with zero out-degree (no outgoing connecting edges), where $0 \leq n \leq N$ and $0 \leq m \leq M$. A sub-pathway is then defined as a path starting from start-node S_n and terminating at end-node D_m. The search for paths between pairs (S_n, D_m) can yield one or more sub-pathways, since multiple sub-pathways may share a start and/or an end node (as shown in Fig. 3.3). Thus, by traversing G between each start and end node, a set S of sub-pathways will be extracted. If no end node is visited within a specific number of steps, the algorithm backtracks and searches another possible path which may lead to an end node.

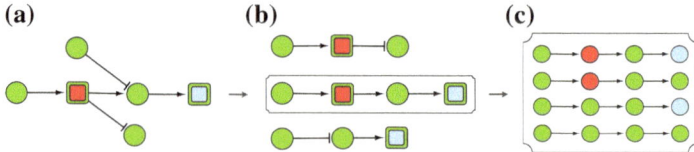

Fig. 3.3 Extraction of linear sub-pathways. **a** A pathway network with two start nodes and two end nodes. *Single circles* denote a gene/gene product, while *overlapping squares* denote multiple gene/gene products per node. **b** Three possible linear sub-pathways extracted from the pathway network, where some nodes correspond to multiple genes/gene products. **c** Expansion of the second sub-pathway to four sub-pathways where all nodes correspond to a single gene/gene product

However, the number of sub-pathways in KEGG pathway graphs may reach the order of billions, increasing the time complexity of both the extraction process and meta-processing analysis. A common approach for avoiding the unwanted increase in complexity is to employ the following steps: (i) exploit the default grouping of genes in KEGG pathway maps—a typical KEGG pathway graph contains inter-actions between nodes corresponding to groups of genes, which are frequently large. Thus sub-pathway extraction happens in two phases. Initially, a set S_c of compact sub-pathways is extracted from G, whose nodes also consist of groups of genes (examples shown in Figs. 3.3, 3.4). Subsequently, each compact sub-pathway is further expanded to a set S of sub-pathways, each consisting of

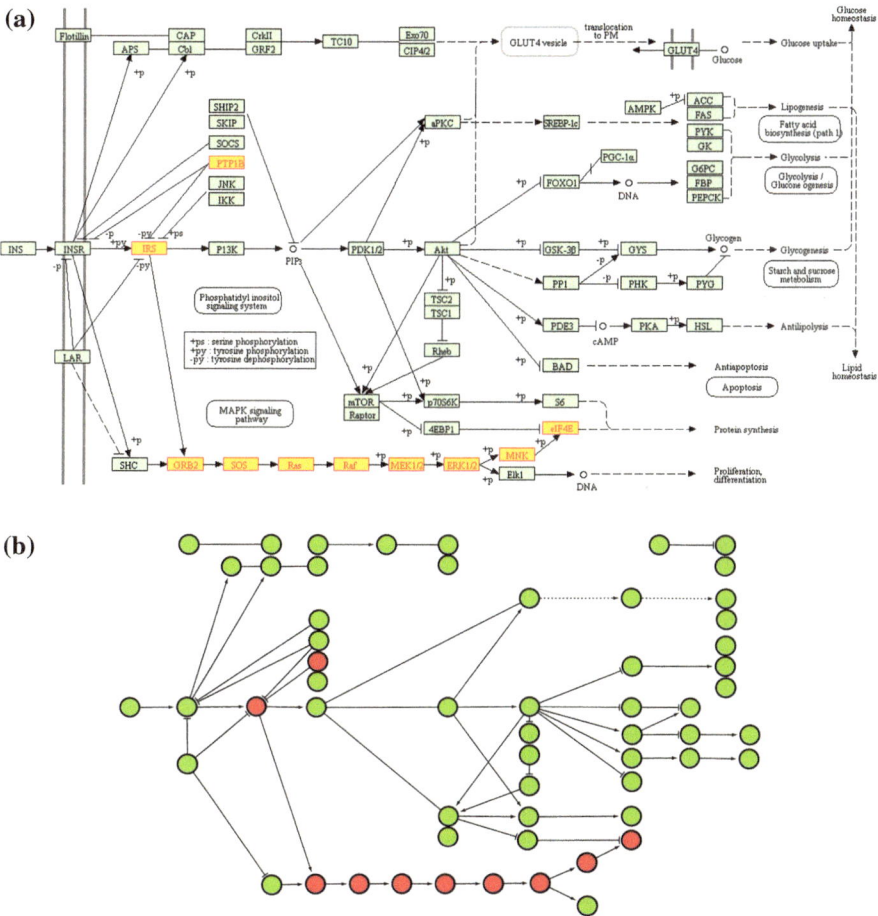

Fig. 3.4 a Insulin signaling pathway from KEGG, with a compact linear sub-pathway of length 10 highlighted. Several sub-pathway member contain more than one members. For instance, the fifth member, *Ras*, corresponds to HRAS, KRAS, and NRAS. **b** The equivalent pathway graph for the insulin signaling pathway

single genes per node (example shown in Fig. 3.3). (ii) Some restrictions are imposed in order to prevent the extraction of extremely long sub-pathways. Extracted sub-pathways may be arbitrarily long, which generally does not reflect the biological reality. Thus, expanding long compact sub-pathways with oversized group nodes may result in an exponential increase in the number of extracted sub-pathways, which in turn induces bias on any statistical meta-processing used to attribute biological significance to the retrieved sub-pathways. To this regard, published methods examine multiple sub-pathway ranges varying from a handful of genes to the border of experimental practicality, and compare their statistics over several organisms. Subsequently, thresholds on the number of connecting edges are imposed, such that minimal loss of information occurs. For example, in CHRONOS, the authors perform sub-pathway statistics for the three most common organisms: *Homo sapiens*, *Mus musculus,* and *Rattus norvegicus*. They observe that the vast majority of the organism's genes are present in extracted sub-pathways ranging from three to ten genes; this range is sufficient to tackle the inherent complexity of pathways without loss of valuable information. Other similar steps to reduce the search complexity for sub-pathways are implemented also by TimeTP (which imposes a shortest possible delay on their interacting genes cross-correlation measure) and TRAP (which introduce a fixed time lag factor on their perturbation factor measure), as described below in Sect. 3.4.

3.3.2 Nonlinear Sub-pathway Extraction

The identification of nonlinear sub-pathways in the form of cliques is highly relevant for network-based analysis in molecular biology. Cliques are maximally connected functional modules whose members are functionally related and have been shown to be highly co-expressed (Matsunaga et al. 2009). Nonlinear sub-pathways are generally extracted by using the k-clique algorithm on pathway graphs which have been converted to an undirected graph (Fig. 3.5). The k-clique algorithm can be summarized as follows: Let D be the adjacency matrix of G and $D(i, j)$ be the shortest path from node i to node j. Johnson's algorithm is used to fill D; let $N = max(D(i, j)$ for all i,

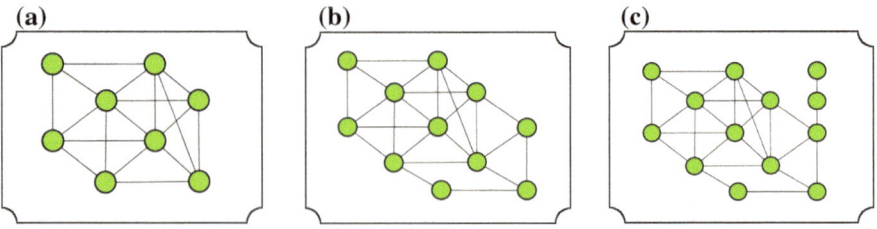

Fig. 3.5 Nonlinear sub-pathways. **a** 3-clique. **b** 4-clique. **c** 5-clique. The distance between any two nodes is no greater than 3, 4 or 5 respectively

j; (2) each edge is a 1-clique by itself; (3) for $k = 2, ..., N$, try to expand each $(k-1)$-clique to *k*-clique: (3.1) consider a $(k-1)$-clique the current *k*-clique K_C; (3.2) repeat the following: if for all nodes *j* in K_C, $D(v, j) \leq k$, add node *v* to K_C; (3.3) eliminate duplicates; (4) the whole graph is a *N*-clique.

In CHRONOS, the optimal sub-pathway length range in nonlinear cascades is obtained based on the same restrictions described in Sect. 3.3.1. The authors suggest that sub-pathways with length up to ten members cover more than 70% of each organism pathway map, and that the entire pathway maps are fully covered by sub-pathways with length up to 20 members. However, for such lengths dozens of interactions arise and thus, in CHRONOS the sub-pathway length is limited to 10 genes, when searching for cliques. The default setting for parameter *k* is set to 2 but users have the possibility to adjust *it*.

In timeClip, Martini et al., follow a procedure previously implemented in *clipper* (Martini et al. 2013; Massa et al. 2010) by which, the search for cliques within pathways is preceded by the moralization and triangulation of the pathway graph. Moralization inserts an undirected edge between two nodes that have a child in common and then removes directions on the edges, and triangulation inserts edges in the moralized graph, so that all cycles of size ≥ 4 have chords (chords being defined as edges connecting two nonadjacent nodes of a cycle). This procedure is done in timeClip in order to assist the search for junction trees (paths of cliques) within the pathway graph.

3.4 Temporal Dynamics Scoring Schemes

3.4.1 *Approach Followed in CHRONOS*

The identification of time-varying portions of pathways relies on various node or interacting edge level metrics or measures, which are usually based on gene expression and/or pathway topology information. For example, the core component of CHRONOS is formed by two scoring schemes, which manage to encapsulate the fold changes in the expression of interacting genes and the temporal aspects describing connecting edge dynamics, as well as the flow of information imposed by the pathway topology.

In the context of pathway networks, the change in the expression level of genes is estimated at the level of edges connecting interacting genes by considering the fold change activity (relative to a control condition). To accomplish this, CHRONOS utilizes the fold change interactivity (FCI) score (Eq. 3.1), which is based on two multivariate logistic functions. This enables the identification of pairs of nodes with high absolute fold change values and with high positive or negative correlation (Kim et al. 2011). The fold activity score of an interaction *e* at time *t* for two connected gene nodes *i, j* is defined as:

$$\text{FCI}_e^t = \text{sgn}\left(f_i^t, f_j^t\right) \cdot A$$

$$A = \left(1 + c \cdot \sum_{k=(i,j)} e^{-K\left(|f_k^t| - T\right)}\right)^{-1} - \left(1 + c \cdot \sum_{k=(i,j)} e^{-K\left(-|f_k^t| - T\right)}\right)^{-1}, \tag{3.1}$$

where f_i^t and f_j^t are the log$_2$-fold change values of nodes i and j, respectively. K and c are the parameters controlling the shape of the multivariate logistic distribution, and T is a shifting parameter.

Furthermore, in CHRONOS, a model rooted in Markov processes is employed for analyzing the temporal dynamics among interacting gene nodes, based on their expression changes, which is able to capture the time-varying aspects of interactions (Jethava et al. 2011). To accomplish this, the time-varying interactivity (TVI) score is defined, which arises from a probabilistic generative model.

Briefly, for a pathway graph G with V nodes (genes) and E interacting edges $[G = (V, E)]$, x_v^t is the fold change expression level for gene $v \in V$ at time t and w_e^t is the interaction strength of $e \in E$ at time t. Weight w can take values in the range $w = \{-2; -1; 0; 1; 2\}$ and represents the interaction strength. Higher values of w correspond to increasing degrees of positive correlation of the interacting genes along the respective edge. The probability of the fold change expression levels for genes i and j at time t, conditioned on the interaction strength w_e^t, can therefore be estimated as:

$$P\left(\overrightarrow{X}^t = \vec{x}^t \mid W^t = \vec{w}^t\right) = \frac{1}{Z(\vec{w}^t)} \exp\left(\sum_{e=(i,j) \in E} w_e^t x_i^t x_j^t\right), \tag{3.2}$$

where $Z(\vec{w}^t)$ is the normalization constant. The edge weights are assumed to evolve according to the Markov property, and thus the probability of transition from state $\vec{w}^{t-1}$ at time $t - 1$ to state $\vec{w}^t$ at time t can be estimated as:

$$Q\left(\vec{w}^{(t-1)}, \vec{w}^t\right) = \frac{1}{Z(\vec{w}^t)} \exp\left(\sum_{e=(i,j) \in E} w_e^t x_i^t x_j^t\right)$$
$$P\left(\overrightarrow{X}^{t-1} = \vec{x}^{t-1} \mid W^{t-1} = \vec{w}^{t-1}\right) \tag{3.3}$$

Based on the above equations, a general score taking into consideration the temporal dynamics underlying each interaction $e \in E$ at each time t can be defined as: $\text{TVI}_e^t = \max_w Q\left(\vec{w}^{t-1}, \vec{w}^t\right)$. Interactions e with scores $\text{TVI}_e^t = \{-2, 0, 2\}$ indicate that the fold changes of the corresponding genes between times $t - 1$ and t, exhibit a strong negative correlation, are uncorrelated and have a strong positive correlation, respectively. Similarly, interactions having scores $\text{TVI}_e^t = \{-1, 1\}$ indicate weak negative and weak positive correlations, respectively.

The two scoring schemes defined above are subsequently merged into a unified score, termed the interactivity score (*IS*), after taking into account the KEGG pathway information on the type of relation of the interaction under consideration (activation or inhibition). Specifically, two consecutive criteria must be satisfied before unifying the FCI and TVI scores. First, the two scores must agree in terms of sign. Second, once the first criterion is fulfilled, the sign of the KEGG type of relation [(+) for 'activation' and (−) for 'inhibition'] must agree with the sign of the two scores. If above described criteria are met, the two scores are multiplied by two weighting factors and then added [e.g. IF FCI > 0 AND TVI > 0 AND relation type = '+' THEN IS is calculated based on Eq. (3.4) below]; otherwise, if one of the two criteria are not met, then *IS* is set to zero (e.g., IF FCI < 0 AND TVI > 0 THEN IS = 0, IF FCI < 0 AND TVI < 0 AND relation type = '+' THEN IS = 0). In case interactions with KEGG relation type 'unknown' are encountered, the other criteria are not checked since the relation type cannot be unambiguously determined. For interaction e at time t, IS is calculated as:

$$IS_e^t = a \cdot \left| TVI_e^t \right| + b \cdot \left| FCI_e^t \right| \tag{3.4}$$

where a and b are weighting parameters indicating the degree of contribution of TVI and FCI score, respectively. These criteria promote interactions among nodes which simultaneously change in terms of expression at each time point and whose variations are in accordance with the constraints imposed by the topology.

Subsequently, at pathway level, the IS can be used to evaluate whole sub-pathways and rank them accordingly. Thus, a sub-pathway score (*subscore*) is calculated as the summand of all IS scores within the sub-pathway and normalized by the total number of interactions in the sub-pathway. For a sub-pathway with N gene members and $N - 1$ interactions at time t, *subscore* is computed as:

$$subscore^{(t)} = \frac{\sum_{e=1}^{N-1} IS_e^t}{N - 1} \tag{3.5}$$

where IS_e^t is the interactivity score of interaction e at time t. This approach allows CHRONOS to identify context- and time-specific sub-pathways, whose genes exhibit correlated expression changes in specific time points and which are in accordance with the flow of information imposed by the sub-pathway topology.

Furthermore, a sub-pathway's enrichment in any user-defined set of 'interesting genes' (e.g., list of differentially expressed genes (DEGs), list of disease-related genes, list of drug target genes, etc.) can be examined, via the cumulative hypergeometric distribution. Specifically, the statistical significance is calculated based on Eq. (3.6) where S is the user-defined gene set, D the number of interesting genes in the user defined set, l the total number of genes in the sub-pathway and d the number of interesting genes found in the sub-pathway under study.

$$P = 1 - \sum_{n=0}^{d} \frac{\binom{D}{n}\binom{S-D}{l-n}}{\binom{S}{l}} \tag{3.6}$$

The obtained P-values are subsequently corrected for multiple testing resulting in FDR corrected q-values. In CHRONOS sub-pathways with $subscore >0.4$ in at least one time point and q-value <0.05 (in case a set of 'interesting genes' is provided) are considered enriched. The $subscore$ threshold is determined by setting a FDR of 1% cutoff on the results derived from 100 permutations; in each permutation a small fraction of sub-pathway interaction scores changed and sub-pathway scores were recomputed (Allantaz et al. 2012).

CHRONOS is also able to identify miRNA regulators of sub-pathways though a simple procedure which blends with the rest of the methodology. Briefly, a micronome layer is constructed above each sub-pathway (Fig. 3.6), based on available miRNA databases, such as miRecords and TarBase, by searching for miRNAs known to target each sub-pathways gene members. A miRNA is considered as a potential regulator of a sub-pathway if its gene targets are significantly enriched in the relevant sub-pathway members, based on the hypergeometric test (Eq. 3.6).

Additionally, at pathway level, CHRONOS provides information about the overall connectivity of the sub-pathway gene members with respect to the complete pathway network. Specifically, the sub-pathway degree ($subDEG$) and betweenness centrality ($subBC$) measures are defined in order to capture the local and global structural aspects of sub-pathway interconnections. Furthermore, the $subPathness$ measure is defined to examine the functional aspects of pathway topologies. When zooming out from sub-pathways to the organism map, it is preferable to view pathways as modules, defined not only through structural measures, but also based on biological criteria which characterize the mechanisms between pathways. Under this notion, $subPathness$ quantifies the degree to which a sub-pathway serves as a

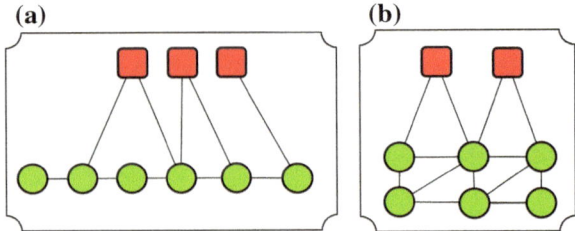

Fig. 3.6 The micronome layer over a sub-pathway. Genes are depicted as *green circle*, while microRNAs as *red rounded rectangles*. Each gene may be regulated by more than one microRNA, while one microRNA may regulate more than one genes. **a** A linear sub-pathway regulated by three microRNAs. **b** A nonlinear sub-pathway regulated by two microRNAs

bridge between different pathways of an organism (Dimitrakopoulou et al. 2014; Kovacs et al. 2010). The *Pathness* of a gene/node i in P pathway maps is:

$$Pathness\,(i) = \sum_{a=1}^{P} \sum_{b=1,b\neq a}^{P} T\,(a,b,i),\qquad(3.7)$$

where $T(a, b, i)$ is the area-overlap between pathway graphs a and b including node i. Generalizing at the level of sub-pathways, the mean *Pathness* of all genes/nodes that belong to the same sub-pathway can define a new measure termed *subPathness*:

$$subPathness\,(i) = \frac{1}{N} \cdot \sum_{j=1}^{N} Pathness\,(j),\qquad(3.8)$$

where N is the number of genes/nodes belonging to sub-pathway i.

3.4.2 Other Approaches

In TimeTP, the initial pathway network consists of gene nodes to which time vectors $\vec{v}$ are assigned, representing the digitized expression values of genes in a differential form, namely 1 (over-expressed), -1 (under-expressed) and 0 otherwise. For example, if the expression of a gene is measured at T time points and has control and treatment conditions, values -1, 1 or 0 will be assigned to each time point with respect to the differential expression (with respect to control) digitized vector $\vec{v}$ of length T. If the data are generated in a single condition, the digitized expression vector $\vec{v}$ contains, at each time point, a value computed as the difference relative to the first time point, or as the difference in expression values of consecutive time points) resulting in a vector of length $T - 1$ (Jo et al. 2016).

For each pathway, TimeTP identifies the time-varying sub-pathway by searching for valid edges among the edges in the original pathway graph. The edges which are time dependent are determined by examining the relationship between differential expression vectors of their connected nodes. A two-step criterion is followed: first, every edge of the time-varying sub-pathway is required to propagate the differential expression pattern along the direction indicated in the original KEGG information. Let there be an edge $N_1 \to N_2$ from a node N_1 to a node N_2, with differential expression vectors $\vec{v}_1$ and $\vec{v}_2$, respectively. Cross-correlation can then be used to estimate the direction of propagation and the amount of delayed time points for the pair of expression vectors $\vec{v}_1$ and $\vec{v}_2$:

$$(\vec{v}_1 * \vec{v}_2)(n) = \sum_{t=-\infty}^{\infty} \vec{v}_1(t)\, \vec{v}_2(t+n), \tag{3.9}$$

where $\vec{v}(t) = 0$ for $t \leq 0$ or $t > T$. Cross-correlation is maximized where the two vectors overlap most, depending on the lag parameter n (delay of n time points). Through this simple measure TimeTP is able to find the shortest possible delay n between two differential expression vectors $\vec{v}_1$ and $\vec{v}_2$:

$$d(\vec{v}_1, \vec{v}_2) = \mathrm{argmax}_n \; (\vec{v}_1 * \vec{v}_2)\,(n) \tag{3.10}$$

When the value $d(\vec{v}_1, \vec{v}_2)$ of a directed edge $N_1 \rightarrow N_2$ is negative, the direction of the expression propagation contradicts the original direction indicated by KEGG and, thus, the edge is considered as invalid and excluded from the analyzed sub-pathway.

In a second step, additional edges with a long positive delay are filtered out, by imposing a threshold for delay, so that the expression propagation in the sub-pathway is bounded within a user-defined time period. Time-varying sub-pathways with one edge are disregarded. The significance of the retrieved time-dependent sub-pathways is estimated by the permutation test. The null hypothesis is that sub-pathways determined by TimeTP are randomly generated and, to test the hypothesis, differential expression vectors for all gene are randomly reassigned and sub-pathways are sampled according to the same procedure described above. Since the ratio of DEGs is not abnormally high, resampled sub-pathways determined following the permutation procedure are most likely to have a significantly shorter path length. Cross-correlation values of edges contained within a certain sub-pathway are likely to be smaller, and therefore, a sum of the cross-correlation of every node pair in the sampled sub-pathway can be chosen as pathway-level statistic.

Time Series RNA-seq Analysis Package (TRAP) extends two classical static pathway analysis algorithms, ORA and SPIA, for the analysis of time series RNA-seq data (Jo et al. 2014). TRAP adapts the perturbation factor (PF) measure described in SPIA to account for temporal dynamics of expression data. Originally, PF is defined by its authors as a gene-level statistic which describes the potential of a gene to relay a perturbation signal along a pathway (Tarca et al. 2009). PF is estimated as a sum of a gene's measured change in expression and a linear function of the perturbation factors of all genes in the pathway directly upstream from the scrutinized gene. In TRAP, authors define the log fold change of expression values measured with RNA-seq experiments as $\Delta E_t(g) = \log\left(Y_t(g)\,/\,X(g)\right)$ where t denotes time points. The major difference between original SPIA and time-series SPIA described by TRAP is in the formula of estimating PF. Due to its static nature, the original SPIA assumes that all downstream genes are affected by the upstream genes at the same time point. In TRAP, however, downstream genes both at the current time point and at the next time point are affected by upstream genes, in a Markov-like process, as defined in Eq. 3.11 below. A time-lag factor α is used to adjust the effect of previous and current upstream gene. $PF(g)$ and $\Delta E_t(g)$ are subsequently defined as the summation of the respective measure over all available time points to derive overall measures of time-variance for individual genes g, as shown in Eq. 3.12:

$$PF_t(g) = \Delta E_t(g) + \sum_{g_u \in US} \beta \cdot \left(\alpha \frac{PF_{t-1}(g_u)}{N_{DS}(g_u)} + (1 - \alpha) \frac{PF_t(g_u)}{N_{DS}(g_u)} \right) \qquad (3.11)$$

$$PF(g) = \sum_{t=1}^{T} PF_t(g) \quad \text{and} \quad \Delta E(g) = \sum_{t=1}^{T} \Delta E_t(g), \qquad (3.12)$$

where g_u are the upstream genes of g, β is a parameter indicating the type of interaction (1 for activation, -1 for inhibition), N_{DS} is the number of downstream genes of g, and US is the set of all upstream genes of g. Generally, one can assume the time-lag factor α is 1, which imposes the restriction that the PF of downstream genes are affected only by the previous time points.

Through this simple extension of SPIA, TRAP can estimate the time-dependent perturbation effect and capture the flow of interactions. Additionally, the net perturbation accumulation at the level of each gene, Acc, can be computed as the difference between the perturbation factor PF of a gene and its observed expression change, as below:

$$Acc(g) = PF(g) - \Delta E(g) \qquad (3.13)$$

Subsequently, the total net accumulated perturbation for all genes i in a sub-pathway or pathway can then be computed as:

$$t_A = \sum_i Acc(g_i) \qquad (3.14)$$

Sub-pathway significance can be estimated with the help of two independent probability values P_{NDE} and P_{PERT}. $P_{NDE} = P(X \geq N_{DE}|H_0)$, captures the significance of the given pathway (or sub-pathway) P_i as estimated by an over-representation analysis (based on the hypergeometric test) of the number of differentially expressed genes (N_{DE}) observed on the pathway. The null hypothesis states that the genes that appear as differentially expressed on a given pathway (sub-pathway) are completely random. The second probability P_{PERT} describes the probability of observing a total accumulated perturbation T_A for the whole pathway (sub-pathway), greater than t_A just by chance: $P_{PERT} = P(T_A \geq t_A|H_0)$.

In timeClip, authors also adapt a static pathway analysis method, *clipper* (Martini et al. 2013), to account for time-varying portions of pathways. To achieve this, they follow a two-step procedure which employs PCA and regression first at pathway level, to reduce dimensionality, and subsequently at sub-pathway level (Martini et al. 2014). In the first step, the first principal component (PC) found by applying the PCA along the temporal dimension of the gene expression data matrix, is explored for temporal variation. The pathways which are identified as time-dependent are decomposed in the second step, into a junction tree which highlights the portion mostly dependent on time.

Briefly, if $X_{n \times t}$ is the gene expression matrix with genes on the rows and time points t on the columns, let $X_{p \times t}^P$ be the submatrix of genes belonging to pathway P, which consists of p genes. Then, PCA is applied to the transpose of X^P, and $Z_{p \times t}^P$ is the resulting scores matrix while $L_{p \times t}^P$ is the resulting loadings matrix. The resulting PCs are stored in vectors $Z_1^P, \ldots, Z_p^P$. The first PCs, which capture the largest proportion of variance in the data, also summarize most of the temporal variation of the genes in pathway P, and are referred to as $Z_i^P(t)$. Then, time series PCs $Z_i^P(t)$ can be decomposed as $Z(t) = p(t) + \in(t)$, where $p(t)$ is a deterministic function, called 'trend,' and $\in(t)$ is the realization of a stationary stochastic process with mean zero. It is assumed that $\in(t)$ follows a continuous time Gaussian autoregressive process of order 1.

A reasonable choice for fitting through regression the trend component is a polynomial of degree 2 in t, resulting in:

$$p(t) = \beta_0 + \beta_1 t + \beta_2 t^2, \tag{3.15}$$

with β_1 capturing existing temporal behaviors of $Z_i^P(t)$ and β_2 correcting for potential nonlinearities. The significance of pathways is expressed by means of the P-value of the test of nullity of β_1 (obtained by a t-test adjusted for multiple hypotheses testing using the Bonferroni correction).

In the second step, pathways identified as time dependent in step 1 are then moralized, triangulated and decomposed into a junction tree following the procedure proposed by *clipper* (Martini et al. 2013). Junction tree algorithm is then used for constructing hyper-trees having cliques as nodes and satisfying the running intersection property, meaning, for any cliques C_1 and C_2 in the tree, every clique on the path connecting C_1 and C_2 contains $C_1 \cap C_2$ (Martini et al. 2014).

For each clique k of pathway P, noted as C_k^P (with $k = 1, \ldots, K$), and composed by a subset of genes in P, c_k^P, the sub-matrix X_{ck}^P of X contains the expressions of genes of the clique C_k^P. Subsequently, for each clique k of P the same procedure described in step 1 is followed: transformation by PCA and then a linear model fitting with polynomial trend and autoregressive process of order 1 on the first PCs. The individual P-Value of clique k, $p_{C_k^P}$ is given by the p of the β_1 resulting from the polynomial regression. Finally, the most significant time-dependent sub-paths along the junction trees within a pathway P, are identified as in (Martini et al. 2013). Briefly, a path is a defined as a chain of consecutive cliques exhibiting significant time dependence (with $p_{C_k^P} \leq 0.05$) with gaps at most of size one. The relevance of each sub-path $j = 1, \ldots, J$, is computed based on the weights of each clique i along the sub-path. First, a measure S_{ij} is defined as:

$$S_{ij} = \sum_{k=1}^{i} \delta_{kj}, \quad \text{with} \quad i = 1, \ldots, Lj, \tag{3.16}$$

where

$$\delta_{kj} = \begin{cases} -\log\left(pC_k^j\right), & \text{if} \quad pC_k^j \leq 0.05 \\ \log\left(1 - pC_k^j\right) & \text{if} \quad pC_k^j > 0.05 \end{cases} \tag{3.17}$$

Then, the overall relevance of sub-pathway j is defined as the normalized maximum of S_{ij}, which can also be used to compare sub-pathways of different lengths:

$$SR_j = \frac{\max_i(SR_{ij})}{L_j} \tag{3.18}$$

Thus, for each time-dependent pathway, timeClip returns a list of relevant paths, ranked according to their relevance.

3.5 Synthetic and Biological Data Analysis Results

Typically, the accuracy of time-varying pathway and sub-pathway analysis methods is tested using both synthetically created and real biological datasets. Statistical significance of the retrieved time-dependent sub-pathways is assessed using permutation tests, usually by randomly reassigning gene expression values to the genes in the dataset followed by resampling of sub-pathways, as described in Sects. 3.4.1 and 3.4.2. Various statistical tests, most commonly the hypergeometric test followed by correction for multiple testing, can also be performed to test for enrichment of biologically relevant information within the retrieved sub-pathways. Results can be easily visualized using various graph-based tools available for pathway analysis methods (e.g., identified active sub-pathways can be highlighted within KEGG maps), as well as tools able to highlight temporal dynamics, such as wheel of time plots, or circular diagrams.

Synthetic pathway networks and synthetic expression data can be generated following a simple procedure: random graphs are created as biological pathway networks by using two closely related graph models, the Erdős–Renyi and Edgar Gilbert. Synthetic gene expression values for each time point can then be generated, such that genes follow a time-dependent up/down regulation, or have no change over time with respect to a control state ($\log_2$-fold change difference 1, -1 and 0, respectively). In addition, the up/down regulated gene expression profiles can be categorized according to the time extent of their regulation.

An important feature that sub-pathway analysis methods must exhibit, in order to highlight the relevance of their results, is the node coverage, i.e., their ability to exploit the maximum number of nodes present in pathway maps. CHRONOS, TEAK, and Subpathway-GM share a similar approach for converting KEGG KGML pathway files into biologically concise gene–gene networks, as described in Sect. 3.2. The authors of CHRONOS examine the node coverage of the extracted sub-pathways in both real and synthetic networks based on the human KEGG pathway maps and Erdos–Renyi and Edgar Gilbert model graphs, respectively, and show that CHRONOS outperforms the other two approaches in both real and synthetic networks with statistical significance (two-sided Wilcoxon signed rank test, $P < 0.01$).

A more difficult task is the evaluation of methods performance in capturing time-varying 'themes.' This is most commonly done using the synthetic datasets, as in their case the experimenter holds the ground truth information on time-dependent regulation. A comparison approach followed in CHRONOS is to apply all methods on a number of independent synthetic network models and for the top-ranking sub-pathways to be considered for further analysis. To overcome inconsistencies imposed by the different sub-pathway topologies identified by each method, the performance was assessed in terms of precision, recall, and F1-score, based solely on the unique nodes included in the top-ranking sub-pathways. The up-/ down-regulated and nonregulated genes included in final results can be defined as TP and FPs, respectively. Up-/down-regulated genes belonging to the corresponding perturbed pathways which were not included in the final sub-pathways can be considered as FNs.

When investigating the IFN-g mRNA and miRNA data (Nazarov et al. 2013) using CHRONOS, a significant re-ranking of sub-pathways was observed during time evolution. The experimental data are two real time series mRNA and miRNA datasets obtained from experiments performed after stimulation of human A375 melanoma cell line with IFN-g (data collected at 0, 3, 12, 24, 48, and 72 h). In more detail, fourth time step corresponding to the experiment performed at 48 h is identified as the peak of activity in the timeline of (miRNA-mediated) non-metabolic sub-pathway frequency distribution. This observation is in agreement with the findings of the original experimental study, which reported no significant alteration of the mRNA levels after 48 h (Nazarov et al. 2013). Furthermore, when evaluating the sub-pathways retrieved by CHRONOS in terms of the structural and functional measures described in Sect. 3.4 (*subDEG*, *subBC*, and *subPathness*), at the 48 h time point the majority of enriched sub-pathways acquired high scores indicating that at this time point a large number of sub-pathways turn into hubs and bridges, which in turn results in fast scattering and propagation of the signal to multiple signaling pathways.

The advantages of using temporal dynamic methods are pinpointed when investigating experimental data which exhibits high time-dependent variability of the pathway functional properties. In such cases crucial biological processes may be neglected or washed out across the whole time period by classic static pathway analysis methods. Generally, time-varying pathway analysis methods have the explanatory power to offer focused serial temporal snapshots of the mechanisms perturbed under specific conditions with resolution of up to single time points. Such methods offer holistic views of the dynamics of molecules participating in multiple sub-pathways and in many cases multiple pathways.

References

Allantaz F, Cheng DT, Bergauer T et al (2012) Expression profiling of human immune cell subsets identifies miRNA-mRNA regulatory relationships correlated with cell type specific expression. PLoS ONE 7:e29979

Büchel F, Rodriguez N, Swainston N et al (2013) Path2Models: large-scale generation of computational models from biochemical pathway maps. BMC Syst Biol 7:1

Cicek AE, Qi X, Cakmak A et al (2014) An online system for metabolic network analysis. Database bau091

Chen X, Xu J, Huang B et al (2011) A sub-pathway-based approach for identifying drug response principal network. Bioinformatics 27:649–654

Dimitrakopoulou K, Dimitrakopoulos GN, Sgarbas KN, Bezerianos A (2014) Tamoxifen integromics and personalized medicine: dynamic modular transformations underpinning response to tamoxifen in breast cancer treatment. OMICS 18:15–33

Gelius-Dietrich G, Desouki AA, Fritzemeier CJ, Lercher MJ (2013) Sybil–efficient constraint-based modelling in R. BMC Syst Biol 7:125

Goeman JJ, Bühlmann P (2007) Analyzing gene expression data in terms of gene sets: methodological issues. Bioinformatics 23:980–987

Haynes WA, Higdon R, Stanberry L, Collins D, Kolker E (2013) Differential expression analysis for pathways. PLoS Comput Biol 9:e1002967

Jacob L, Neuvial P, Dudoit S (2012) More power via graph-structured tests for differential expression of gene networks. Ann Appl Stat 561–600

Jethava V, Bhattacharyya C, Dubhashi D, Vemuri GN (2011) Netgem: network embedded temporal generative model for gene expression data. BMC Bioinformatics 12:1

Jo K, Kwon HB, Kim S (2014) Time-series RNA-seq analysis package (TRAP) and its application to the analysis of rice, Oryza sativa L. ssp. Japonica, upon drought stress. Methods 67:364–372

Jo K, Jung I, Moon JH, Kim S (2016) Influence maximization in time bounded network identifies transcription factors regulating perturbed pathways. Bioinformatics 32:i128–i136

Judeh T, Johnson C, Kumar A, Zhu D (2013) TEAK: topology enrichment analysis framework for detecting activated biological subpathways. Nucleic Acids Res 41:1425–1437

Kanehisa M, Sato Y, Kawashima M, Furumichi M, Tanabe M (2015) KEGG as a reference resource for gene and protein annotation. Nucleic Acids Res gkv1070

Kim Y, Kim TK, Kim Y et al (2011) Principal network analysis: identification of subnetworks representing major dynamics using gene expression data. Bioinformatics 27:391–398

Kovács IA, Palotai R, Szalay MS, Csermely P (2010) Community landscapes: an integrative approach to determine overlapping network module hierarchy, identify key nodes and predict network dynamics. PLoS ONE 5:e12528

Li C, Han J, Shang D et al (2012) Identifying disease related sub-pathways for analysis of genome-wide association studies. Gene 503:101–109

Li C, Han J, Yao Q et al (2013) Subpathway-GM: identification of metabolic subpathways via joint power of interesting genes and metabolites and their topologies within pathways. Nucleic Acids Res 41:e101

Li X, Shen L, Shang X, Liu W (2015) Subpathway analysis based on signaling-pathway impact analysis of signaling pathway. PLoS ONE 10:e0132813

Martini P, Sales G, Massa MS, Chiogna M, Romualdi C (2013) Along signal paths: an empirical gene set approach exploiting pathway topology. Nucleic Acids Res 41:e19

Martini P, Sales G, Calura E, Cagnin S, Chiogna M, Romualdi C (2014) timeClip: pathway analysis for time course data without replicates. BMC Bioinformatics 15:1

Massa MS, Chiogna M, Romualdi C (2010) Gene set analysis exploiting the topology of a pathway. BMC Syst Biol 4:1

Matsunaga T, Yonemori C, Tomita E, Muramatsu M (2009) Clique-based data mining for related genes in a biomedical database. BMC Bioinformatics 10:1

Moutselos K, Kanaris I, Chatziioannou A, Maglogiannis I, Kolisis FN (2009) KEGGconverter: a tool for the in-silico modelling of metabolic networks of the KEGG Pathways database. BMC Bioinformatics 10:1

Nam S, Chang HR, Kim KT et al (2014) PATHOME: an algorithm for accurately detecting differentially expressed subpathways. Oncogene 33:4941–4951

Nazarov PV, Reinsbach SE, Muller A, Nicot N, Philippidou D, Vallar L, Kreis S (2013) Interplay of microRNAs, transcription factors and target genes: linking dynamic expression changes to function. Nucleic Acids Res 41:2817–2831

Sales G, Calura E, Cavalieri D, Romualdi C (2012) Graphite—a bioconductor package to convert pathway topology to gene network. BMC Bioinform 13:1

Sales G, Calura E, Martini P, Romualdi C (2013) Graphite Web: web tool for gene set analysis exploiting pathway topology. Nucleic Acids Res gkt386

Sebastian-Leon P, Vidal E, Minguez P et al (2014) Understanding disease mechanisms with models of signaling pathway activities. BMC Syst Biol 8:1

Subramanian A, Tamayo P, Mootha VK et al (2005) Gene set enrichment analysis: a knowledge-based approach for interpreting genome-wide expression profiles. In: Proceedings of the national academy of sciences 102:15545–15550

Tarca AL, Draghici S, Khatri P et al (2009) A novel signaling pathway impact analysis. Bioinformatics 25:75–82

Vrahatis AG, Dimitrakopoulou K, Balomenos P, Tsakalidis AK, Bezerianos A (2016) CHRONOS: a time-varying method for microRNA-mediated subpathway enrichment analysis. Bioinformatics 32:884–892

Wrzodek C, Büchel F, Ruff M, Dräger A, Zell A (2013) Precise generation of systems biology models from KEGG pathways. BMC Syst Biol 7:1

Chapter 4
Identification of Differentially Expressed Pathways and Sub-pathways

Abstract This chapter provides a survey of pathway and sub-pathway-based differential expression analysis. Differential expression analysis, the comparison of genes' expression across various conditions, has become the primary tool for finding biomarkers, drug targets, and understanding the molecular mechanisms of disease. Within this context, a recent trend is that of developing pathway topology-based methods, which integrate the benefits of gene set-based analysis and augment them with prior information on the underlying gene interactions from pathway databases. It is discussed how approaches based exclusively on gene set analysis ignore the position of genes within pathways and therefore may miss relevant information regarding the differential activation of pathways in certain conditions. In turn, pathway-based methods better account for cases when a pathway may be activated by the significant expression of a single gene (such as a cellular membrane receptor), which may, in turn, significantly perturb downstream genes. Additionally, pathway-based methods are able to model cases in which the differential expression of several downstream genes may not have the same effect on the whole pathway if the upstream receptor gene is not activated. The chapter briefly presents gene set-based methods, and subsequently overviews various aspects related to topology-based pathway analysis methods: conversion of pathway database information to graphs, the use of gene and pathway-level statistics highlighting differential expression, and the evaluation of statistical significance of differentially expressed sub-pathways. Finally, a tool for the identification of differentially expressed sub-pathways is presented as case study.

Keywords Pathway analysis · Sub-pathways · Differentially expressed genes · Differential expression analysis · Differentially expressed pathways · Over-representation analysis · Functional class scoring · Statistical significance · Pathway topology · Pathway scoring

© The Author(s) 2017
A. Bezerianos et al., *Computational Methods for Processing and Analysis of Biological Pathways*, SpringerBriefs in Computer Science, DOI 10.1007/978-3-319-53868-6_4

4.1 Introduction

Differential expression analysis, the comparison of genes' expression across various conditions, has become the primary tool for finding biomarkers, drug targets, and understanding the molecular mechanisms of disease. As highlighted in Chap. 1, the first approaches in differential expression analysis typically relied on simple gene-centric approaches, where statistical tests and/or correlation analysis were used to select candidate genes exhibiting differential expression between treatment and control samples. Subsequently, various approaches based on gene sets were introduced, to alleviate the obvious loss of biological context characterizing the aforementioned approaches. Gene set analysis methods consider sets of genes simultaneously when performing differential expression analysis, thus accounting for the complex association relationships among genes, especially in the context of achieving biological function. The advantages of gene set analysis are that it allows researchers to characterize groups of genes functioning within the same pathways and, consequently, it eases the identification of active pathways enriched with sets of genes differentially expressed between various conditions.

A recent trend in differential expression analysis is that of developing pathway topology-based methods, which integrate the benefits of gene set-based analysis, and augmenting them with existing information on gene interactions from pathway databases. Approaches based exclusively on gene set analysis ignore the position of genes within pathways and therefore may miss relevant information regarding the differential activation of pathways in certain conditions. As an example, a pathway may be activated by the significant expression of a single gene (e.g., cellular membrane receptor), which may, in turn, significantly perturb downstream genes, whereas solely the differential expression of several downstream genes may not have the same effect on the whole pathway if the upstream receptor gene is not activated (Tarca et al. 2009). Section 4.2 of this chapter briefly describes gene set-based methods, and subsequently, Sects. 4.3–4.6 present in detail various aspects related to topology-based methods.

4.2 Approaches Based on Gene Sets

As already discussed in the Chap. 1, methods for gene set analysis can be categorized as *Over-Representation Analysis* (ORA) and *Functional Class Scoring* (FCS) (Khatri et al. 2012). Briefly, methods based on the ORA approach calculate pathway significance by estimating the probability of observing a number of differentially expressed genes in a given pathway by chance alone, using the hypergeometric and chi-square statistical tests. The genes differential expression is commonly assessed using a threshold on fold change of expression. One of the most popular early ORA-based methods is the Database for Annotation, Visualization and Integrated Discovery (DAVID) (Dennis et al. 2003), which provides an extensive set of data mining and visualization tools. FCS methods overcome the need for using arbitrary expression thresholds to determine differential expression, on which ORA methods are based. Instead they produce ranked lists of genes resulting from correlation with phenotype and statistical tests for differential expression. One of the first and most prominent among FCS methods is the *Gene Set Enrichment Analysis* (GSEA) (Subramanian et al. 2005). GSEA first ranks all background genes according to their (differential) expression, commonly correlated with phenotype distinction, and then computes an enrichment score for a certain gene set under study. This is done based on the distribution of the gene set toward the top or bottom of the whole ranked genes list. The enrichment score is estimated by walking down the ranked gene list and incrementing the running sum score whenever a gene present in the gene set in encountered and decrementing it otherwise. The maximum value of the running sum is chosen as final score of the gene set:

$$\mathrm{ES}(S) = \max_i |P_{\mathrm{hit}}(S, i) - P_{\mathrm{miss}}(S, i)|, \qquad (4.1)$$

where

$$P_{\mathrm{hit}}(S, i) = \sum_{\substack{g_j \in S \\ j \le i}} \frac{|r_j|^p}{N_R}, \quad P_{\mathrm{miss}}(S, i) = \sum_{\substack{g_j \notin S \\ j \le i}} \frac{1}{(N - N_H)},$$

with r_j being the correlation of the expression profile of gene j with a phenotype or profile of interest, S is the gene set under scrutiny with N_H genes out of the total N genes in the experiment. Parameter p is used for weighting the genes (usually $p = 1$, when genes are weighted based on their correlations), while P_{hit} and P_{miss} denote the fraction of genes in S present or absent, respectively, in the ranked list up to position i.

The main difference between the ORA and FCS methods is that, while ORA relies on the selection of a subset of differentially expressed genes based on expression threshold criteria, subsequently checked for enrichment within various pathways, FCS considers the entire set of measured genes and computes first a gene-level statistic based on differential expression of individual genes. Then, FCS methods aggregate the gene-level statistic into a pathway-level one, which may be multivariate (and thus account for inter-dependencies among genes) or univariate. Finally, FCS methods assess the statistical significance of the pathway-level statistic, based on one of two types of null hypotheses: competitive (in which gene labels for each gene set—pathway—are permuted and then genes in the pathway are compared with genes that are not in the pathway), and self-contained (in which class labels of samples—conditions—are permuted and then the set of genes within the pathway is compared with itself) (Khatri et al. 2012).

However, ignoring the complex gene interactions structure within pathways diminishes the relevance of analysis, obscuring the presence of important biological signals. Gene set-based approaches, such as ORA and FCS methods, consider only the number of genes contained in the pathways when performing the differential expression analysis, disregarding valuable information of pathway structure present in curated databases. Therefore, as long as the set of genes under study is the same, ORA and FCS methods will produce the same results, even if the underlying structure of gene interactions is totally different (Khatri et al. 2012).

4.3 Approaches Based on Pathway Topology

The majority of topology-based methods developed for detecting differentially expressed pathways (or sub-pathways) follows a sequence of steps that is partially overlapping with those employed by time-varying methods described in Chap. 3: (i) conversion of pathway databases information into graphs (networks) which allow subsequent computational modeling (described in Sect. 4.4), (ii) contextual-ization of the transformed pathway graphs using expression data, or metabolites profiling data, in the case of methods working with metabolic networks, (iii) search within graphs for paths or network motifs corresponding to portions of pathways which are activated by treatment or disease, step which is commonly accompanied by (iv) the use of various gene or interaction-level measures or statistics which are usually further aggregated into pathway-level measures, and (v) testing the statistical significance of the pathway-level measure (all described in Sect. 4.5).

One of the first pathway-based differential expression analysis method is SPIA (Tarca et al. 2009). SPIA combines aspects of the two types of gene set-based approaches (ORA and FCS) into an approach which incorporates prior knowledge with pathway topology. It does so by assessing the perturbations caused on a pathway by the changes in gene expression across the whole of the pathway. It considers both the over-representation of differentially expressed genes among the pathway genes and the perturbations propagated across the pathway topology. To achieve this, it relies on two independent probability measures P_{NDE} and P_{PERT}.

Probability P_{NDE} is defined as the probability that at least N_{DE} differentially expressed genes are present on a pathway:

$$P_{NDE} = P(X \geq N_{de}|H_0), \qquad (4.2)$$

where H_0 is the null hypothesis employed by the model, stating that the genes present in a pathway appear purely by chance and that the phenotype has no correlation with them. Probability P_{NDE} follows the hypergeometric distribution:

$$P(X = i) = \frac{\binom{m}{i}\binom{n}{k-i}}{\binom{m+n}{k}}, \quad x \in 0, \ldots, k, \qquad (4.3)$$

where m is the number of genes within the pathway, n is the number of genes not present within the pathway, and k is the number of differentially expressed genes.

Probability P_{PERT} assesses the extent of the perturbation on a pathway and is quantified through a gene perturbation factor:

$$\text{PF}(g_i) = \Delta E(g_i) + \sum_{j=1}^{n} \beta_{ij} \cdot \frac{\text{PF}(g_j)}{N_{\text{ds}}(g_j)}. \tag{4.4}$$

The first term represents the fold change of gene g_i, while the second represents the sum of perturbation factors of its upstream genes, normalized by the number of its downstream genes. Coefficients β_{ij} quantify the interaction strength between genes g_i and g_j, with positive values denoting activating and negative values denoting inhibitory relations between the two genes. The sum of all equations describing the perturbation for all genes can be summarized as follows:

$$\text{PF} = \Delta E + B \cdot \text{PF},$$

$$B = \begin{bmatrix} \frac{\beta_{11}}{N_{\text{ds}}(g_1)} & \cdots & \frac{\beta_{1n}}{N_{\text{ds}}(g_n)} \\ \vdots & \ddots & \vdots \\ \frac{\beta_{n1}}{N_{\text{ds}}(g_1)} & \cdots & \frac{\beta_{nn}}{N_{\text{ds}}(g_n)} \end{bmatrix},$$

$$\Delta E = \begin{bmatrix} \Delta E(g_1) \\ \vdots \\ \Delta E(g_n) \end{bmatrix}, \quad \text{PF} = \begin{bmatrix} \text{PF}(g_1) \\ \vdots \\ \text{PF}(g_n) \end{bmatrix} \tag{4.5}$$

The net accumulation of perturbations for each gene can be denoted as the difference between the perturbation factor and the fold change:

$$\text{Acc}(g_i) = \text{PF}(g_i) - \Delta E(g_i) \tag{4.6}$$

P_{PERT} then shows the probability that a perturbation T_A is more significant than the sum of the net accumulation of perturbations for all genes on the pathway, and is calculated by checking whether $t_A = \sum_i \text{Acc}(g_i)$ appears as a result of the phenotype or purely by chance:

$$P_{\text{PERT}} = P\left(T_A \geq \sum_i \text{Acc}(g_i)|H_0\right) \tag{4.7}$$

To this end, T_A is computed for a number of random scenarios, where in each one, $N_{\text{DE}}(P_i)$ genes are selected from the pathway P_i randomly and are considered as differentially expressed. In estimating P_{PERT} (Tarca et al. 2009) use a bootstrap approach based on random permutations of differentially expressed genes ID labels.

Probabilities P_{NDE} and P_{PERT} are ultimately merged into a single probability P_G used to rank pathways according to their relevance to the observed phenotype and to check the hypothesis that the pathway is significantly perturbed due to a specific phenotype:

$$P_G = c_i - c_i \cdot \ln(c_i), \quad c_i = P_{\mathrm{NDE}}(i) \cdot P_{\mathrm{PERT}}(i). \tag{4.8}$$

The time-varying TRAP method (Jo et al. 2014) described in Chap. 3 is based on the approach followed by SPIA.

The Subpathway-GM method (Li et al. 2013) was devised to identify biologically significant metabolic sub-pathways by incorporating gene expression information and metabolite measurements with pathway information. Initially, the method matches 'genes of interest' and 'metabolites of interest' from a specific study with the enzymes and metabolites corresponding to the nodes of a metabolic pathway network, characterized as signature nodes. Next, it searches for similarities in the distances of the signature nodes within the pathway topology, in order to extract biologically significant sub-pathways. Namely, this procedures create sub-pathways by allowing a certain number of nonsignature nodes on the paths connecting signature nodes. The statistical significance of these sub-pathways is then assessed using a hypergeometric test.

In Subpathway-GM, KEGG metabolic pathways are used as a source of prior knowledge for constructing the background pathway graph. Each pathway is converted to a directed graph according to the biochemical reactions described at the corresponding KGML file, as presented in Sect. 3.2. The nodes of the resulting network represent enzymes and metabolites. If a metabolite participates as a substrate or a product in a reaction, an edge connects the node corresponding to the metabolite and the node corresponding to the enzyme. In the case of an irreversible reaction, the direction of the edge is from the substrate to the enzyme. If the reaction is reversible, the edge is undirected.

For each network containing nodes of interest, the shortest path is calculated between any two nodes. If this path has length less that $n + 1$, where n is the maximum number of nodes allowed between two nodes of interest, then the nodes are added to a node-set. Next, for each of those node-sets, the sub-network corresponding to the node-set is extracted from the original pathway network. Sub-networks with more than s nodes are considered as sub-pathways. Parameter s should be selected in a way to portray that smaller sub-networks cannot correspond to biological processes. The statistical significance of the sub-pathways is assessed using a hypergeometric test:

$$P = 1 - \sum_{x=0}^{r_g + r_m - 1} \frac{\dbinom{t_g + t_m}{x} \dbinom{m_g + m_m - t_g - t_m}{n_g + n_m - x}}{\dbinom{m_g + m_m}{n_g + n_m}}, \tag{4.9}$$

where m_g is the number of background genes, t_g the number of the sub-pathway genes, n_g the number of genes of interest and r_g the number of gene of interest within the sub-pathway. Symbols m_m, t_m, n_m, r_m correspond to the same quantities for the metabolites.

The TEAK method (Judeh et al. 2013) was developed to detect sub-pathways activated by underlying biological processes. Initially, the method constructs a directed network for each KEGG pathway, and subsequently, based on gene expression data, it extracts linear and nonlinear sub-pathways from each network. Identified sub-pathways are then scored using the Bayesian Information Criterion (BIC) for context-specific studies and the Kullback-Leibler (KL) divergence for case-control studies.

In order to extract a linear sub-pathway, the method accesses the network through root nodes and traverses the network, marking successively each node as visited, until it reaches a node without any outgoing edges (leaf node). The start node, the intermediate nodes and the end node correspond to a sub-pathway. Next, it backtracks on that sub-pathway, marking the nodes as not visited, until it finds the first node offering an alternate path to reach an end node. The original start node, the originally visited nodes, the newly visited nodes, and the end node correspond to a second sub-pathway. This process is repeated until all sub-pathways connecting the start node to all possible end node are extracted, and for all start-nodes of the network.

In order to extract a nonlinear sub-pathway, TEAK identifies communities of cliques focusing on one important feature: the extracted cliques are directed feed-forward loops of size three (Palla et al. 2005). Additionally, their method allows for gene nodes to participate in multiple sub-pathways and due to the choice of feed-forward loops, no sub-pathway contains nodes, or links, whose removal would separate the sub-pathway.

If the experimental data are context-specific, the method fits a Bayesian network for each sub-pathway and utilizes the Bayesian information criterion to score it:

$$score_{\text{BIC}} = \log P(D|\hat{\theta}) - 0.5d\log N, \tag{4.10}$$

where D denotes the gene expression data, N the number of data samples, d the number of parameters used to represent a linear Gaussian node, and $\hat{\theta}$ the approximation of the maximum likelihood estimate of the parameters. A node G connected to m upstream nodes $G_{j=1...m}$ is linear Gaussian if

$$P(G|G_1, \ldots, G_m) = N(\beta_0 + \beta_1 G_1 + \cdots + \beta_m G_m; \sigma^2), \tag{4.11}$$

where $\beta_0,\ldots,\beta_m$ denote the regression coefficients and σ^2 the variance. Each node is scored separately and the sub-pathway score is the normalized sum of the gene member scores.

If the experimental data are case-control, the method adjusts a Bayesian network for each of the case and control data, transforms them to an equivalent Gaussian form and calculates the Kullback-Leibler divergence between the two networks q and p:

$$\text{KL}(q \parallel p) = \frac{1}{2}\log\frac{|\Sigma_p|}{|\Sigma_q|} + \frac{1}{2}\text{Tr}\left(\Sigma_p^{-1}\Sigma_q\right) + \frac{1}{2}\left(\mu_q - \mu_p\right)^{\text{T}}\Sigma_p^{-1}\left(\mu_q - \mu_p\right) - \frac{h}{2},$$

$$(4.12)$$

where μ is the vector of medians, Σ the covariance matrix, $|\Sigma|$ it's determinant, Tr the trace of a matrix, and h the number of nodes in each network.

The DEAP method (Haynes et al. 2013) can be used to identify the most differentially expressed linear sub-pathway (path) in a biological pathway through a relatively simple recursive procedure. The method maps the gene expression data on a pathway network which it accepts as input. Next, it calculates the differential expression of the linear sub-pathway by adding the gene expression value of a successive node which is connected with an edge corresponding to an 'activation' event, and subtracting the gene expression value of a successive node which is connected with an edge corresponding to an 'inhibition' event. The sub-pathway is recursively constructed and scored by starting from its last edge:

$$score_N = B_N; \; score_i = B_i + T_{i-1,i} \cdot score_{i-1}, \quad i = N - 1, \ldots, 1, \qquad (4.13)$$

where $T_{i-1,i} \in \{-1, 1\}$ depending on whether the edge connecting node $i - 1$ and i corresponds to an inhibition, or activation event, respectively. The method performs the previous steps for every edge of the network and eventually returns the maximum absolute score, as well as the sub-pathway which achieves it.

In order to test the significance of the retrieved sub-pathway, the null hypothesis on its expression is tested using a random rotation sampling approach, in which the original data (genes) is repeatedly multiplied by a rotation matrix, until a null distribution is generated. The rotation procedure handles correlation structure within gene sets by conditioning the rotations on the covariance matrix. This is achieved by performing first an orthogonal projection of the original data on the residual space from a linear model, then rotate, and subsequently transform the rotated data back (Dorum et al. 2009).

Thus, to approximate a null distribution of the test statistics, n rotations of the data are performed. For each rotation sample, the DEAP score, $score_i$, is computed. The P-value is calculated as a proportion of simulated scores that are greater or equal than the score $score^*$ of the extracted sub-pathway:

$$p = \frac{\#(score_i \geq score^*)}{n} \qquad (4.14)$$

The random rotation procedure has the advantage of producing reasonable results without the need for large number of samples, as in the case of permutation tests.

The Clipper method (Martini et al. 2013) employs a different approach in order to identify sub-pathways which exhibit significant correlations with a given phenotype. The two-step approach first reconstructs a pathway network utilizing a junction tree structure consisting of cliques (Cowell 1998). Next, it extracts linear sub-pathways between the root and a leaf of each junction tree, scores them, and assesses their statistical significance. Several pathway databases are used as sources of prior knowledge, namely Biocarta (Nishimura 2001), KEGG (Kanehisa et al. 2015), NCI/Nature Pathway Interaction Database (Schaefer et al. 2009), and Reactome (Joshi-Tope et al. 2005). The pathways of each database are converted to a gene interaction network for each pathway using the approach of graphite (Sales et al. 2012), and described in Sect. 3.2.

Each directed pathway network, which may be either acyclic or cyclic, is converted to an equivalent undirected acyclic graph by utilizing the junction tree structure as follows from (Cowell 1998): first, a single undirected edge is added to the network, connecting any two nodes sharing a downstream node (moralization), creating an undirected moral graph G. The term moralization is used since any two parents are eventually connected if they share a child. Next, additional edges are added to G, such that all cycles of length ≥ 4 have chords, i.e., edges connecting nonneighboring nodes of a cycle. This process is known as triangulation. Finally, all cliques of the resulting graph are identified and a junction tree is constructed. Any clique which lies on the path connecting two cliques C_1 and C_2, contains the nodes of the node-set $C_1 \cap C_2$.

Thus, each node of the junction tree corresponds to a clique. Each clique is assigned a weight w which is equal to the P-value of the homoscedasticity test. To this end, the Breusch-Godfrey test (Breusch 1978; Godfrey 1978) is used with two Gaussian models as input, one for the 'case' samples and a second for the 'control' expression data corresponding to the genes in clique C:

$$\mathbf{M}_i(C) = \left\{ Y \sim N_P(\mu_i, \Sigma_i), K_i = \Sigma_i^{-1} \in S^+(C) \right\}, i = 1, 2, \qquad (4.15)$$

where P is the number of genes, K_i is the inverse of the covariance matrix for the Gaussian model, and $S^+(C)$ the set of positive definite matrices with null elements corresponding to the missing edges of clique C. The P-Value denotes the statistical significance of the clique across experimental conditions, with $P < 0.05$ denoting significant cliques.

Subsequently, all possible sub-pathways starting from the root of a junction tree and ending at a leaf are extracted. Each sub-pathway is a cascade of statistically significant cliques which may contain no more than one nonsignificant clique. Let j be the sub-pathway, L_j the sub-pathway length, w_{kj} the weight of the kth clique in sub-pathway j, and m the position of the clique with maximum score within the sub-pathway. The biological significance of the sub-pathway is assessed using the following scoring scheme, which is similar to the approach followed by timeClip (Martini et al. 2014) and described in Sect. 3.4. Namely

$$SR_j = \max_i \left(\sum_{k=1}^{i} \delta_{kj} \right) \cdot \frac{m}{L_j}, \quad i = 1, \ldots, L_j, \tag{4.16}$$

$$\delta_{kj} = \begin{cases} -\log(w_{kj}), \, w_{kj} \le 0.05 \\ \log(1 - w_{kj}), \, w_{kj} > 0.05 \end{cases}. \tag{4.17}$$

For each pathway, the sub-pathway with a maximal SR is selected as its representative. Finally, sub-pathways exhibiting significant overlaps between them are removed.

Another recent method, which was developed to work for a small sample dataset, is the Pathway and Transcriptome information (PATHOME) (Nam et al. 2014). In contrast with other methods, such as SPIA and Clipper, PATHOME does not rely on a permutation based approach and/or a large number of samples to obtain a null distribution for the statistical test. In a first step, PATHOME decomposes the pathways into linear sub-pathways (paths) from leaf nodes to root nodes which comply a predefined association rule, and subsequently uses simple statistical tests to evaluate the significance of differential expression patterns along the path in terms of consecutive nodes' correlations.

Briefly, based on prior knowledge on pathway structure extracted from KEGG, a depth-first search (DFS) algorithm is used to exhaustively decompose pathway maps into a large number of linear sub-pathways. A reduced set of candidate sub-pathways is created based on an association rule between the regulation information of interacting genes and their expression correlation, in terms of the sign of the Pearson product-moment correlation coefficients. Namely, KEGG interactions of type 'activation' must be confirmed by positive correlation between the expression values of their corresponding genes in the experimental data. Similarly, in case of interactions type 'inhibition', the expression correlation between the respective genes is assumed to be negative, in order for the sub-pathway containing the respective interaction to be considered valid. The procedure is followed separately for each experimental group. A sub-pathway is expanded upstream, starting from the leaf node, by successively adding interactions satisfying the association rule, and stops progressing when the rule ceases to be satisfied. Each sub-pathway is then compared between the k experimental groups (usually $k = 2$, for data from treatment and control experiments) and if its length in each of the group is greater than three, it is kept for subsequent statistical significance analysis. The length of a sub-pathway in the kth group can be expressed as

$$l^k = \operatorname{argmin}_m \left\{ -\sum_{i=1}^{m} I\left(\operatorname{sgn}\left(r^k_{i,i+1} \times e_{i,i+1}\right) = 1\right) + \sum_{i=1}^{m} R\left(\operatorname{sgn}\left(r^k_{i,i+1} \times e_{i,i+1}\right)\right) \right\} + 1 \tag{4.18}$$

$$\text{with: } R(x) = \begin{cases} 0, & \text{if } x \in \{1\} \\ \infty, & \text{otherwise} \end{cases},$$

where p is the number of genes in the sub-pathway, $e_{i,i+1}$ is the edge between adjacent genes i, and $i + 1$, $r^k_{i,i+i}$, is the corresponding correlation coefficient between the two genes, sgn($\bullet$) is the sign function, and $I(\bullet)$ is the indicator function. The left term ensures the association rule is satisfied, and allows l^k to progress upstream, while $R(\bullet)$ prevents extension of the sub-pathway when the association rule is not met.

In the test step, sub-pathways are represented as sets of consecutive correlation coefficients $r^k_{i,i+i}$. Fisher transformation is then applied to $\left| r^k_{i,i+i} \right|$ in order to improve their normality approximation:

$$c^k_{i,i+1} = \frac{1}{2} \ln \frac{1 + \left| r^k_{i,i+i} \right|}{1 - \left| r^k_{i,i+i} \right|} \in [0, \infty). \tag{4.19}$$

In cases of sub-pathways of different length in the k conditions, the minimum common length is set to be compared. If μ^k represents the mean of $c^k_{i,i+1}$, for $k = \{1,2\}$, the significance under the null hypothesis is tested: H_0: $\mu^1 = \mu^2$. This is equivalent to suggesting as alternative hypothesis that the global means of sub-pathway expression correlations are different. The authors of (Nam et al. 2014) use a z-test statistic to measure significance, adjusted for multiple comparisons with FDR = 0.05.

4.4   Conversion of Pathway Databases Information to Graphs

All topology-based differential pathway analysis methods have as initial step the creation of a background pathway graph or network, providing prior structural knowledge on the interactions among genes. Only minor differences occur in the way pathway information is merged into networks and in terms of repositories used as sources for the data. Conversion of pathway data from KGML and other formats, such as SBML or BioPax is performed as described in Sect. 3.2.

KEGG is the primary source of pathway information for the majority of methods described above, as well as DEsubs [(Vrahatis et al. 2016b) described in Sect. 4.6]. Exceptions are DEAP, which uses pathway data from the PANTHER database, and Clipper, which besides KEGG, uses also data from Reactome. Some methods use only subsets of KEGG, such as Subpathway-GM (which only uses metabolic pathways) and SPIA (which only uses human signaling pathway). Additionally, the majority of methods create individual graphs for each pathway, the exception being DEsubs, which creates whole organism networks by merging pathways from each organism.

4.5   Gene and Pathway-Level Statistics of Differential Expression

The topology-based methods employed in the analysis of differentially expressed pathways can generally be categorized in two groups: (i) those which make use of experimental data (gene expression, differential metabolite profiles, etc.) embedded within the process of uncovering differentially expressed portions of pathways, and (ii) those which make use of experimental data only in the statistical evaluation of the retrieved sub-pathways.

4.5.1   Methods Using Experimental Data for Sub-pathway Identification

Experimental data are utilized by the majority of topology-based methods in the form of gene expression from microarray experiments (either case-control of context-specific) or, more recently, from RNA-seq experiments, and/or metabolomics experiments. Among the methods in this category, two different types of approaches can be found in current publications: those which use proper expression values in the methodology followed to extract paths (e.g., Subpathway-GM and Clipper) and those using a derived measure (correlation of expression values of two interconnected genes—used in PATHOME).

As described in Sect. 4.3, Subpathway-GM searches for paths across pathway maps containing specific nodes (signature nodes), which are identified as nodes corresponding to differential metabolites and genes found in experimental data. Gaps of as much as two nonsignature nodes are allowed along the paths and a minimum length of 5 nodes is imposed for retrieved paths. Through this procedure the authors ensure the search space extends to incorporate biologically relevant portions of pathways, while, at the same time, constrain the complexity of their search. Statistical significance of the retrieved sub-pathways is assessed by means of an enrichment analysis of user-defined sets of interesting genes, based on the hypergeometric test, as defined by Eq. (4.9).

The approach followed by Clipper broadly resembles the procedure described above, in the sense that in this case paths along differential nodes (cliques) of the junction tree are searched for, with gaps of at most one nondifferential clique being allowed. The difference is in the fact that a statistical significance test is performed here at the level of nodes (cliques) and, thus, embedded into the pathway search procedure. For each clique, its significance in differentiating between two conditions (e.g., treatment and control) is assessed by means of a homoscedasticity test which employs as a null hypothesis the equality in the means of expression values of the tested clique in the two experimental conditions. The resulting P-Value of the test is used as weight for the respective clique in a sub-pathway-level measure summing up the relevance of the respective sub-pathway (see Eqs. 4.16 and 4.17).

PATHOME follows a different approach, in that is uses correlation coefficients between expression values of successive interconnected nodes (based on pathway structure information). The expression correlation measure is embedded into the sub-pathway search procedure and is used as a condition in terminating the search. Specifically, as described in Eq. 4.18, an association rule is defined based on the sign of the correlation coefficient and the information on the edge type derived from KEGG. When the search reaches an interconnected-edge for which the association rule is not satisfied, sub-pathway expansion is terminated. Statistical significance of retrieved sub-pathways is assessed by using sets of Fisher transformed correlation coefficients, corresponding to edges in a sub-pathway, and comparing their means between two experimental conditions under the null hypothesis of having equal means.

4.5.2   Methods Using Experimental Data for Evaluating the Statistical Significance of Sub-pathways

SPIA defines a gene-level measure, termed perturbation factor (PF—see Eq. 4.4), which is subsequently aggregated into a pathway-level measure of accumulated perturbation (t_A). Although PF is defined based on differential expression values and incorporates pathway topology information by including expression of upstream genes and the number of downstream genes, it is ultimately used only for the statistical tests assessing the significance of the differential expression of pathways. As described in Sect. 4.3, two independent statistical tests P_{NDE} and P_{PERT} are merged into a global probability value (Eq. 4.8) employed to rank pathways and test whether a pathway is perturbed as a result of the condition under study.

In TEAK, two different pathway-level measures (BIC and KL) are used for statistical testing and scoring differentially activated sub-pathways. For context-specific data, retrieved linear and nonlinear sub-pathways are fit with a Gaussian Bayesian network, described by conditional probability distributions, as shown in Eq. 4.11. Subsequently, the BIC (Eq. 4.10) is used to score the sub-pathways. This way, both topological and experimental information are used to capture aspects reflecting differentially expressed sub-pathways. For case-control data two different Bayesian networks are fitted, one for each experimental condition. Subsequently, the KL divergence for multivariate Gaussians is used as statistical measure of differential expression (Eq. 4.12).

DEAP defines a much simpler measure for scoring sub-pathways. As described in Eq. 4.13, the sub-pathway level score is constituted by the absolute expression value summated along the nodes within the sub-pathways, considering the signs corresponding to the type of interaction between successive gene nodes. To assess the statistical significance of the retrieved sub-pathways, simulated sub-pathways are created based on a random rotation sampling of gene label identifiers within a pathway, and the null hypothesis of obtaining equally high or higher scores by chance is tested (Eq. 4.14).

4.6   DEsubs: A Flexible Tool for Identification of Differentially Expressed Sub-pathways

To further illustrate the various aspects entailed by the analysis of differentially expressed sub-pathways, this section describes a recent tool which incorporates a wide range of features related to this type of differential analysis. DEsubs is a publicly available tool, part of the Bioconductor project (http://bioconductor.org/packages/DEsubs/). Its use is to extract perturbed sub-pathways by utilizing gene expression data from RNA-seq experiments, which are enriched in topological and/or functional features.

The tool works on RNA-seq expression paired case-control data and utilizes an organism-level pathway network based on KEGG signaling pathway processing described in CHRONOS (Vrahatis et al. 2016a). By constructing organism-level pathway networks, the method sets itself apart from the other topology-based methods described Sect. 4.5, which mostly rely on separate pathway networks constructed from individual KEGG pathway maps. This strategy allows DEsubs to account for overlaps between pathways, a field of study which has attracted significant interest recently (Tegge et al. 2016).

Subsequently, the RNA-seq data are mapped onto the pathway network and two separate approaches are used to extract nodes and interaction edges of interest. First, sets of statistically significant differentially expressed genes (DEGs) are identified based on a user selected tool, which can be chosen from a list of the most common differential expression analysis tools: edgeR (Robinson et al. 2010), DESeq (Anders and Huber 2010), EBSEq (Leng et al. 2013), limma (Smyth 2004), SAMR (Li and Tibshirani 2013), NBPSeq (Di et al. 2011), and TSPM (Auer and Doerge 2011). Each of the tools described above will return a list of statistically significant DEGs, together with their FDR adjusted-values (Q-values), corrected for multiple testing. Alternatively, the user has the option to import a custom ranked list of genes accompanied by their Q-values to proceed with analysis.

Based on the list of DEGs, the nodes V of the original pathway network $G = (V, E)$ are pruned using the Q-value threshold (commonly 0.05):

$$Q - value(i) < Q_{threshold}, \quad i \in V \qquad (4.20)$$

Subsequently, the interacting edges between the genes selected following the *node rule* in Eq. 4.20 are also pruned based on an association rule based on prior biological knowledge on interaction type from KEGG and the expression profiles of interacting genes. This *edge rule* is similar to the one employed by PATHOME and described in Sect. 4.3:

$$\text{cor}(V_i, V_j) * \text{reg}(V_i, V_j) > C_{\text{threshold}}, V_i, V_j \in V \qquad (4.21)$$

If genes V_i and V_j are related with an edge annotated with a KEGG interaction type activation the reg is set to 1, if their interaction type is inhibitory, it is set to -1. The correlation between the expression profiles of the two genes V_i and V_j is calculated using a correlation measure selected by the user between the Pearson, Spearman, or Kendall correlations. In contrast with PATHOME, DEsubs imposes a minimum threshold $C_{\text{threshold}}$ (commonly $C_{\text{threshold}} = 0.6$) on the correlation coefficient values, in order to ensure highly relevant co-expression patterns are retained within the network.

In the next step, sub-pathway extraction is performed based on five main categories: (i) components, (ii) communities, (iii) streams, (iv) neighborhoods, (v) cascades. Each of these sub-pathway types is able to highlight different topological and biological aspect within the network.

There are several types of components that can be searched: regular cliques (in which every two distinct nodes are adjacent), maximal clique (clique with the largest possible number of edges), or k-cores (a maximal sub-pathway in which every node has a degree of at least k). The community category extracts groups of highly interacting genes, based on six different approaches: random walk (finds community structures which minimize the expected description length of a random walker trajectory), walktraps (finds densely connected structures via random walks), modular (finds communities via a modularity measure and a hierarchical approach), leading eigenvector (finds densely connected communities based on the leading nonnegative eigenvector of the modularity matrix), betweenness (detects community structures using edge betweenness), and greedy (detects community structure using greedy optimization of modularity). Stream, neighborhood, and cascade type of sub-pathways are extracted starting from a gene of interest (GOI) node and are able to highlight perturbations paths underlying differentially expressed sub-pathways. Additionally, these types of sub-pathways can be searched either downstream (forward) or upstream (backward) from a certain gene of interest. The component category extracts strongly connected groups of genes (cliques) highlighting dense local areas within the network. These genes may share a common functional property, as mentioned in Sect. 3.3.1.

Candidate GOIs are genes exhibiting specific topological or functional roles within the network. Toward this end, a number of topological measures has been employed from the igraph package (Csardi and Nepusz 2006), capturing both local and global properties of the network: (i) Degree is a local measure which captures the number of direct interactions of a gene. (ii) Betweenness is a global measure which captures the number of shortest paths across the network, passing through a gene. (iii) Closeness is a global measure which is equal to the inverse of the sum of distances from all other genes on the network. (iv) Eccentricity is a global measure which captures the shortest path distance from a gene to the most remote downstream gene. Finally, two additional global measures are employed, namely (v) Kleinberg hub-centrality score and (vi) Google's page rank. Candidate GOIs also include start-nodes, namely genes with no incoming interactions. Alternatively, GOIs can be genes with specific functional roles. Toward this end, a number of gene sets are available to accommodate the analysis: (i) KEGG pathway genes, (ii) biological processes, (iii) cellular components and (iv) molecular functions from Gene-Ontology, disease related genes from (v) OMIM and (vi) GAD, drug targets from (vii) DrugBank, microRNA targets from miRecords, and (viii) transcription factors from Transfac and Jaspar (Barneh et al. 2015; Chen et.al 2013; Li et al. 2011).

Sub-pathways retrieved using the afforementioned methodology can subsequently be further tested for enrichment in certain annotation terms, based on the cumulative hypergeometric distribution, in a manner similar with the one followed by Subpathway-GM:

$$P = 1 - \sum_{x=0}^{d} \frac{\binom{D}{x}\binom{G-D}{l-x}}{\binom{G}{l}}, \tag{4.23}$$

where G is the number of genes in the user-defined input list, l is the number of those genes included in the sub-pathway, D is the number of genes annotated with a specific term, and d is the total number of genes contained in the sub-pathway.

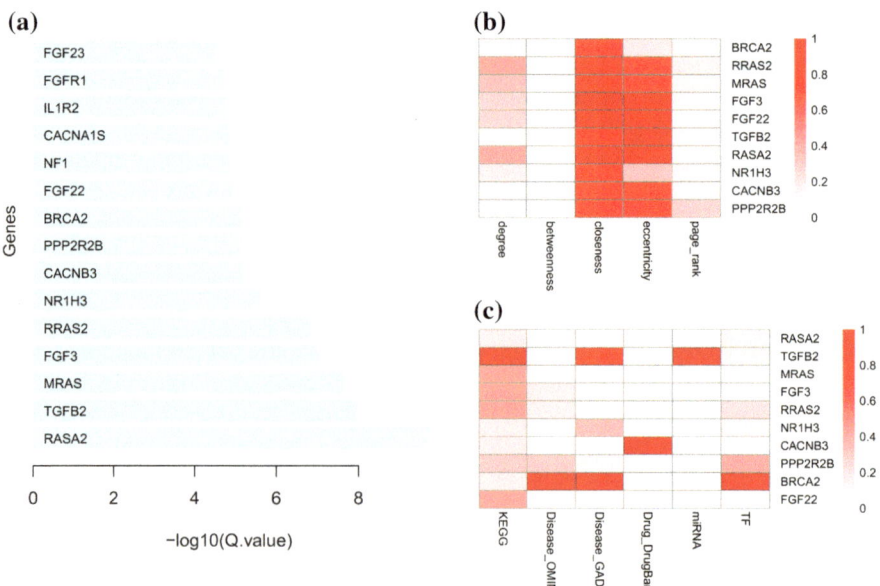

Fig. 4.1 **a** Bar plots illustrating lowest Q-value. **b** Heat-map showing genes with highest topological measures. **c** Heat-map showing genes with highest functional measures (output of DEsubs R-package)

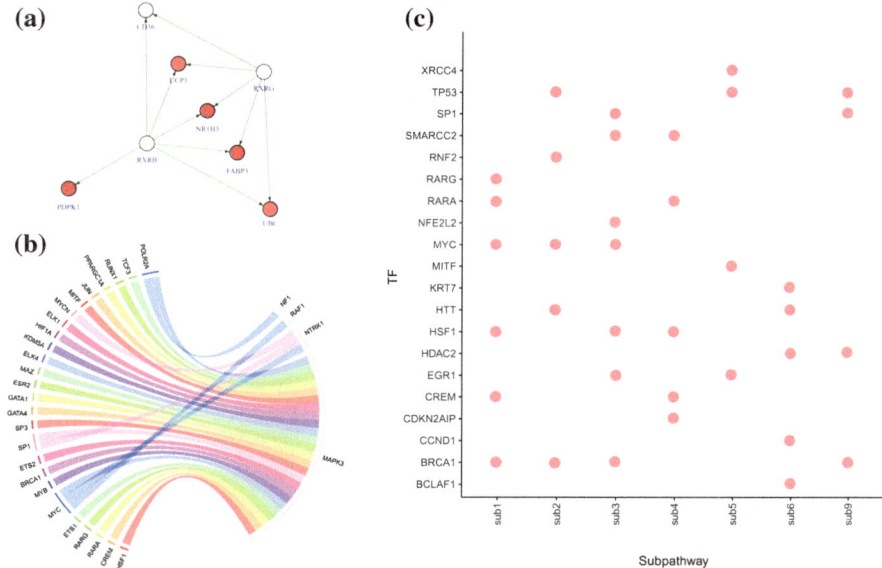

Fig. 4.2 **a** Example of sub-pathway obtained using the community walktrap approach. *Node fill* indicates the *Q*-value, *darker shades*, corresponding to lower values. Edges' *green* or *red* color indicates positive and negative correlation, respectively. **b** Circular diagram displaying association between genes belonging to a sub-pathway and transcription factors. **c** Dot plot diagram showing the enrichment of transcription factors on a set of sub-pathways (output of DEsubs R-package)

Finally, DEsubs enables the visualization of results at different levels: from gene, to sub-pathway and organism. An example of gene-level visualization option is displayed in Fig. 4.1, where the fifteen top genes in terms of *Q*-values, and top genes with highest topological and functional measures are presented. Figure 4.2a displays an example of sub-pathway level visualization, a graph type display of the sub-pathway and the corresponding interactions between contained genes, with nodes and edge color coding for the Q-values and correlation coefficient, respectively. Figure 4.2b presents a circular diagram displaying gene-transcription factor associations within a sub-pathway and Fig. 4.2c shows an example of organism-level visualization providing a global view on KEGG pathway term enrichment over a set of sub-pathways extracted using DEsubs.

References

Anders S, Huber W (2010) Differential expression analysis for sequence count data. Genome Biol 11:1

Auer PL, Doerge RW (2011) A two-stage Poisson model for testing RNA-seq data. Stat Appl Genet Mol Biol 10:1

Breusch TS (1978) Testing for autocorrelation in dynamic linear models. Aust Econ Pap 17:334–355

Barneh F, Jafari M, Mirzaie M (2015) Updates on drug–target network; facilitating polypharmacology and data integration by growth of DrugBank database. Briefings Bioinform bbv094

Chen EY, Tan CM, Kou Y et al (2013) Enrichr: interactive and collaborative HTML5 gene list enrichment analysis tool. BMC Bioinform 14:1

Csardi G, Nepusz T (2006) The igraph software package for complex network research. InterJ, Complex Syst 1695:1–9

Cowell R (1998) Introduction to inference for Bayesian networks. In: Learning in graphical models. Springer, Netherlands, pp 9–26

Dennis G, Sherman BT, Hosack DA, Yang J, Gao W, Lane HC, Lempicki RA (2003) DAVID: database for annotation, visualization, and integrated discovery. Genome Biol 4:1

Di Y, Schafer DW, Cumbie JS, Chang JH (2011) The NBP negative binomial model for assessing differential gene expression from RNA-Seq. Stat Appl Genet Mol Biol 10:1

Dørum G, Snipen L, Solheim M, Sæbø S (2009) Rotation testing in gene set enrichment analysis for small direct comparison experiments. Stat Appl Genet Mol Biol 8:1–24

Godfrey LG (1978) Testing against general autoregressive and moving average error models when the regressors include lagged dependent variables. Econometrica: J Econometric Soc 1293–1301

Haynes WA, Higdon R, Stanberry L, Collins D, Kolker E (2013) Differential expression analysis for pathways. PLoS Comput Biol 9:e1002967

Jo K, Kwon HB, Kim S (2014) Time-series RNA-seq analysis package (TRAP) and its application to the analysis of rice, Oryza sativa L. ssp. Japonica, upon drought stress. Methods 67:364–372

Joshi-Tope G, Gillespie M, Vastrik I et al (2005) Reactome: a knowledgebase of biological pathways. Nucleic Acids Res 33(suppl 1):D428–D432

Judeh T, Johnson C, Kumar A, Zhu D (2013) TEAK: topology enrichment analysis framework for detecting activated biological subpathways. Nucleic Acids Res 41:1425–1437

Kanehisa M, Sato Y, Kawashima M, Furumichi M, Tanabe M (2015) KEGG as a reference resource for gene and protein annotation. Nucleic Acids Res gkv1070

Khatri P, Sirota M, Butte AJ (2012) Ten years of pathway analysis: current approaches and outstanding challenges. PLoS Comput Biol 8:e1002375

Leng N, Dawson JA, Thomson JA et al (2013) EBSeq: an empirical Bayes hierarchical model for inference in RNA-seq experiments. Bioinformatics 29:1035–1043

Li C, Han J, Yao Q et al (2013) Subpathway-GM: identification of metabolic subpathways via joint power of interesting genes and metabolites and their topologies within pathways. Nucleic Acids Res 41:e101–e101

Li J, Tibshirani R (2013) Finding consistent patterns: a nonparametric approach for identifying differential expression in RNA-Seq data. Stat Methods Med Res 22:519–536

Li X, Li C, Shang D et al. (2011) The implications of relationships between human diseases and metabolic subpathways. PloS One 6:e21131

Martini P, Sales G, Massa MS, Chiogna M, Romualdi C (2013) Along signal paths: an empirical gene set approach exploiting pathway topology. Nucleic Acids Res 41:e19–e19

Martini P, Sales G, Calura E, Cagnin S, Chiogna M, Romualdi C (2014) TimeClip: pathway analysis for time course data without replicates. BMC Bioinform 15:1

Nam S, Chang HR, Kim KT et al (2014) PATHOME: an algorithm for accurately detecting differentially expressed subpathways. Oncogene 33:4941–4951

Nishimura D (2001) BioCarta. Biotech Softw Internet Rep Comput Softw J Scient 2:117–120

Palla G, Derényi I, Farkas I, Vicsek T (2005) Uncovering the overlapping community structure of complex networks in nature and society. Nature 435:814–818

Robinson MD, McCarthy DJ, Smyth GK (2010) edgeR: a bioconductor package for differential expression analysis of digital gene expression data. Bioinformatics 26:139–140

Sales G, Calura E, Cavalieri D, Romualdi C (2012) Graphite—a bioconductor package to convert pathway topology to gene network. BMC Bioinform 13:1

Schaefer CF, Anthony K, Krupa S, Buchoff J, Day M, Hannay T, Buetow KH (2009) PID: the pathway interaction database. Nucleic Acids Res 37(suppl 1):D674–D679

Smyth GK (2004) Linear models and empirical bayes methods for assessing differential expression in microarray experiments. Statist Appl Genet Mol Biol 3:1

Subramanian A, Tamayo P, Mootha VK et al (2005) Gene set enrichment analysis: a knowledge-based approach for interpreting genome-wide expression profiles. Proc Natl Acad Sci 102:15545–15550

Tarca AL, Draghici S, Khatri P et al (2009) A novel signaling pathway impact analysis. Bioinformatics 25:75–82

Tegge AN, Sharp N, Murali TM (2016) Xtalk: a path-based approach for identifying crosstalk between signaling pathways. Bioinformatics 32:242–251

Vrahatis AG, Dimitrakopoulou K, Balomenos P, Tsakalidis AK, Bezerianos A (2016a) CHRONOS: a time-varying method for microRNA-mediated subpathway enrichment analysis. Bioinformatics 32:884–892

Vrahatis AG, Balomenos P, Tsakalidis AK, Bezerianos A (2016b) DEsubs: an R package for flexible identification of differentially expressed subpathways using RNA-seq experiments. Bioinformatics btw544